UNIVERSITY OF GLAMORGAN
PRIFYSGOL MORGANNWG

D1348981

TELECOMMUNICATIONS LOCAL NETWORKS

BT Telecommunications Series

The BT Telecommunications Series covers the broad spectrum of telecommunications technology. Volumes are the result of research and development carried out, or funded by, BT, and represent the latest advances in the field.

The series includes volumes on underlying technologies as well as telecommunications. These books will be essential reading for those in research and development in telecommunications, in electronics and in computer science.

1. *Neural Networks for Vision, Speech and Natural Language*
   Edited by R Linggard, D J Myers and C Nightingale.
2. *Audiovisual Telecommunications*
   Edited by N D Kenyon and C Nightingale
3. *Digital Signal Processing in Telecommunications*
   Edited by F A Westall and S F A Ip
4. *Telecommunications Local Networks*
   Edited by W K Ritchie and J R Stern

TELECOMMUNICATIONS LOCAL NETWORKS

Edited by

W K Ritchie

Formerly Principal General Manager responsible for Advanced Telecommunications Research, BT Laboratories, Martlesham Heath, UK

and

J R Stern

Section Manager of Fibre Access Technology, BT Laboratories, Martlesham Heath, UK

CHAPMAN & HALL

London · Glasgow · New York · Tokyo · Melbourne · Madras

Published by Chapman & Hall, 2–6 Boundary Row, London SE1 8HN

Chapman & Hall, 2–6 Boundary Row, London SE1 8HN, UK

Blackie Academic & Professional, Wester Cleddens Road, Bishopbriggs, Glasgow G64 2NZ, UK

Van Nostrand Reinhold Inc., 115 5th Avenue, New York NY10003, USA

Chapman & Hall Japan, Thomson Publishing Japan, Hirakawacho Nemoto Building, 6F, 1–7–11 Hirakawa-cho, Chiyoda-ku, Tokyo 102, Japan

Chapman & Hall Australia, Thomas Nelson Australia, 102 Dodds Street, South Melbourne, Victoria 3205, Australia

Chapman & Hall India, R. Seshadri, 32 Second Main Road, CIT East, Madras 600 035, India

First edition 1993

Printed in Great Britain by St Edmundsbury Press Ltd, Bury St Edmunds, Suffolk

ISBN 0 412 45810 1 0 442 30883 3(USA)

A catalogue record for this book is available from the British Library

Library of Congress Cataloging-in-Publication data available

Contents

Contributors

J Adams	Broadband and Visual Networks, BT Laboratories
P F Adams	Access Networks, BT Laboratories
J Ballance	Access Networks, BT Laboratories
E J Boswell	Access Networks, BT Laboratories
R A Boulter	Digital Services, BT Laboratories
D S Butler	Formerly Access Networks, BT Laboratories
D E A Clarke	Access Networks, BT Laboratories
J W Cook	Access Networks, BT Laboratories
D W Faulkner	Access Networks, BT Laboratories
D I Fordham	Broadband and Visual Networks, BT Laboratories
J R Fox	Access Networks, BT Laboratories
I Gallagher	Broadband and Visual Networks, BT Laboratories
R Guyon	BT Worldwide Networks, London
P J Hawley	System Engineering, BT Laboratories
C E Hoppitt	Network Control Layer, BT Laboratories
S Hornung	Access Networks, BT Laboratories
J D Jenson	Ameritech Services (Science and Technology), USA
K J Maynard	Engineering Network Control, BT Laboratories
T Miki	NTT Transmission Systems, Japan
D I Monro	Formerly Access Networks, BT Laboratories
K A Oakley	BT Worldwide Networks, London
W Rosenau	Deutsche Bundespost Telekom, Germany
J R Stern	Access Networks, BT Laboratories
J Trigger	Access Operations and Maintenance, BT Laboratories
S Worger	Formerly Access Networks, BT Laboratories

Preface

Over the last two decades we have seen revolutionary advances in telecommunications networks. The development of optical fibre and satellite transmission systems have progressively reduced the cost and improved the quality of long-distance and inter-office transmission, while modern switching systems provide near-instantaneous fault-free interconnection; but yet in that last vital kilometre or two from exchange to customer, new technology has been slow to make any significant impact. The local network is still overwhelmingly a copper-pair network employing analogue transmission.

This is not because we do not have the technology — the optical fibre, the lasers, the VLSI circuitry, the signalling systems are all well developed. Rather it is because the local network presents a unique combination of cost/engineering barriers. The embedded investment is huge, while the investment return from domestic customers is small; it is in this environment that the copper pair has provided a very viable (though limited) solution ever since the inception of the telephone. Can modern technology break through these barriers and offer the low-cost network solutions needed to persuade telecommunications companies that there is a sensible case for the huge investment involved in change-over? That primarily is the question, in all its ramifications, addressed in this book.

The challenge of the local network has been a great stimulus to the creative energies of engineers. The point has now been reached where the more attractive network solutions have been explored in depth and are being assessed in field trials in many advanced countries. These developments are showing that the new technologies can indeed offer very substantial advantages in bandwidth, flexibility, longer reach, switch node consolidation, duct space conservation and immunity to electrical interference. These in turn open up a wide range of new service opportunities — services which our modern society will increasingly demand if our businesses are to remain competitive, our private and social lives are to be fulfilled and our environment is to be preserved.

This then is a good time to take stock. Development engineers believe they have largely mapped and evaluated the future possibilities. They can rightly claim that, while there will of course be continuous development, administrators and regulators will shortly have the tecnical/cost information to formulate future policy. There is confidence that the solutions explored will be enduring, at least in basic concept.

The book is broadly in two sections. In the first, leading international workers present their views on local network developments from their national perspective; these have been placed at the beginning to give the reader a wide perspective from which to appreciate the following chapters.

The second section treats each of the main topic areas in depth. The authors are leading experts at the BT Laboratories which has pursued a comprehensive development programme in the local network field over the last 12 years. Of particular note has been the invention and development of passive optical networks (PONS) to the point where they are widely recognized as probably the most promising of network architectures; the development of switched-star networks (SSNs), first as an interactive cable-TV network and later as a fully integrated service network; and the development of digital local loop transmission systems over copper pairs for ISDN where the work has formed the basis for international standards. These highlights are well represented, but they have grown out of a balanced programme which has sought to cover the whole field. The stress is on creative solutions allied to sound, low-cost engineering.

We believe the book presents all the main themes in local network development and gives an in-depth survey at a time when we stand on the threshold of exciting radical change.

W K Ritchie
J R Stern

Part One

National Surveys

1

FIBRE-OPTIC SUBSCRIBER LOOP NETWORKS

T Miki

1.1 INTRODUCTION

Although the era of broadband services is clearly visible on the horizon, there is still no concrete image of exactly what this era will be like. The crucial issues now facing network operators are when and in what manner to make the huge investments necessary to realize broadband ISDN (BISDN) [1,2] and fibre-optic subscriber loop networks in particular.

Over the last few years, quite a large volume of fibre-optic cables and systems for large business customers have been installed, that is to say fibre-to-the-office (FTTO) [3]. From now, the most important task is to decide how to introduce fibre-optic systems to small business and residential customers, that is fibre-to-the-home (FTTH).

While several countries have been proposing fibre-to-the-curb (FTTC) as the best way of creating broadband networks [4], the situation in Japan has led to the belief that FTTH will be successful in the near future. Japan has very high concentrations of customers over wide areas. Most customers are sited right next to a road so the curb space available for optical network units (ONUs) is severely limited and is very costly. Outside maintenance work for ONUs is also expensive and error-prone; it should be eliminated whenever possible. The concept envisaged here is that fully optical subscriber loop networks are indispensable in realizing a successful BISDN; therefore the most cost-effective FTTH solution, from the viewpoint of long-term investment, must be found.

There is no doubt that the technology needed to create a BISDN system — such as optical fibres, very high-speed transmission and switching systems and video band-compression coding methods — is being rapidly developed.

The most promising multiplexing and switching scheme, asynchronous transfer mode (ATM), is quite advanced and capable of constructing a high-performance BISDN. Before BISDN can be realized, however, a very tough conundrum must be resolved. Which comes first — the infrastructure or user demand? This idea is that infrastructure, i.e. optical subscriber loops, must be widely installed well in advance of full BISDN deployment. Network structure modernization is another objective that must be achieved in conjunction with opticalizing the subscriber loop. The structure of conventional local networks is based on the old technologies of metallic loop cables and crossbar local switching systems. These are counter-productive in the fibre optic and digital switching era. Section 1.2 therefore describes the network evolution plans to develop a fibre-optic subscriber network. Section 1.3 considers the services to be offered over the network. Section 1.4 discusses the fibre-optic subscriber systems to be realized within a five year period starting in 1995, as well as cost considerations. Last, the investment strategy is investigated in section 1.5.

1.2 NETWORK EVOLUTION PLAN

NTT has been opticalizing its network since the early 1980s. Following the traditional strategy, short- and long-haul trunk lines were the first to see the introduction of optical fibres. Figure 1.1 gives a fairly complete overview of past practices and future plans towards the introduction of BISDN [5].

Taking into account technological trends, major issues for BISDN include the introduction of fibre-optic subscriber loops and ATM transport techniques. Figure 1.2 shows the three-step scenario for network evolution.

In general, new telecommunications services are initiated for large businesses, and then extended to small businesses and eventually to the home with few exceptions. So far, NTT has been supplying narrowband ISDN and other services over fibre-optic loop systems only to large business customers. ATM leased line services would also be provided just for business customers, for the time being. This piecemeal introduction of fibre-optic systems is uneconomic for smaller business and residential customers and should be replaced in the near future by large-scale introduction which offers many economies of scale. According to our network evolution plan, the introduction of fibre-optic subscriber loop networks should precede the emergence of widespread demand for broadband services. Such demand is expected to arise in the late 1990s from business customers and in or around 2000 from residential customers.

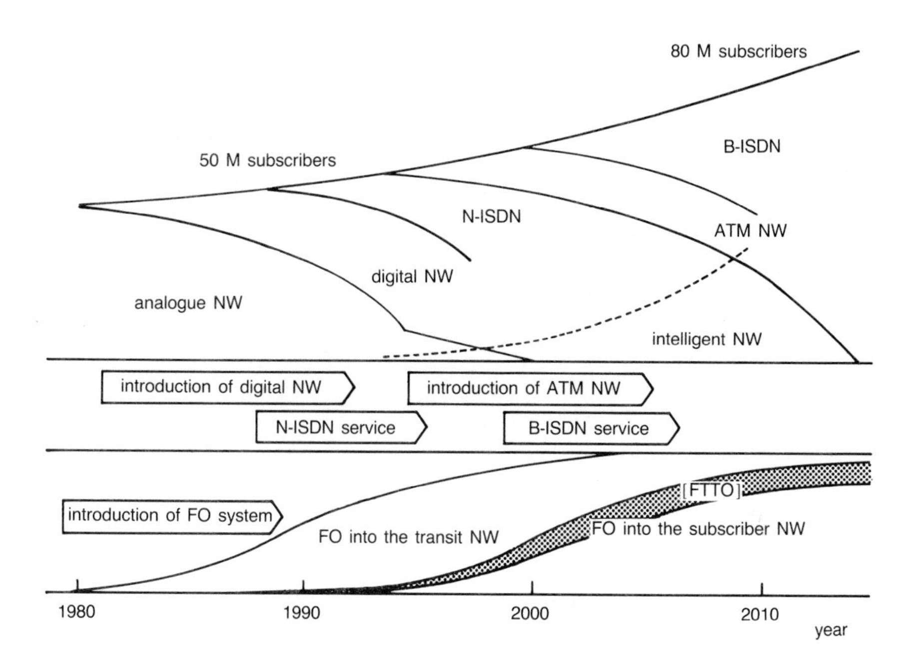

Fig. 1.1 Network evolution towards BISDN and fibre-optic subscriber system introduction plan in NTT.

The introduction of a fibre-optic subscriber network is a golden opportunity to increase dramatically network simplicity and efficiency. NTT aims to evolve its existing network structure [6] as shown in Fig. 1.3. Key points of this evolution are to consolidate the current 7000 local switching offices to around 1000 by increasing the switching capacity of each group centre. Regional and district centres (total 87), where long-distance transit switching systems are located, will be replaced by 54 zone centres. A two-layer switching network architecture will replace the existing five-layer architecture. The main benefit of the two-layer scheme is the significant reduction in cost accompanying the elimination of many manned offices. Another important benefit is the increase in network reliability. Because the future subscriber loop network will use a ring topology, high reliability will be ensured with self-healing network operations [7]. Furthermore, the new subscriber network will adopt double routeing and dual homing schemes for the access network as well as the transfer network.

Both user service and network operation aspects should be considered in subscriber loop network evolution plans. The major features of the proposed subscriber loop network are listed in Fig. 1.4.

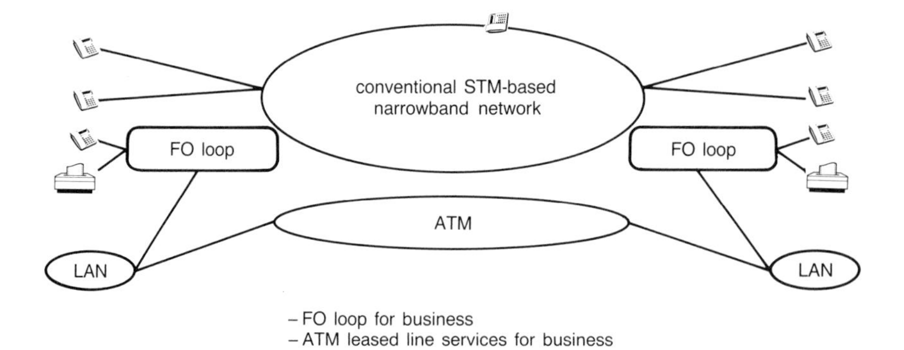

early 1990s networks

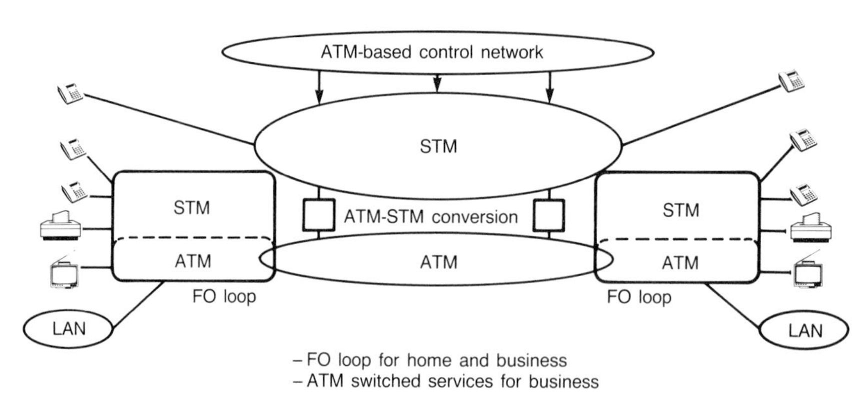

late 1990s networks

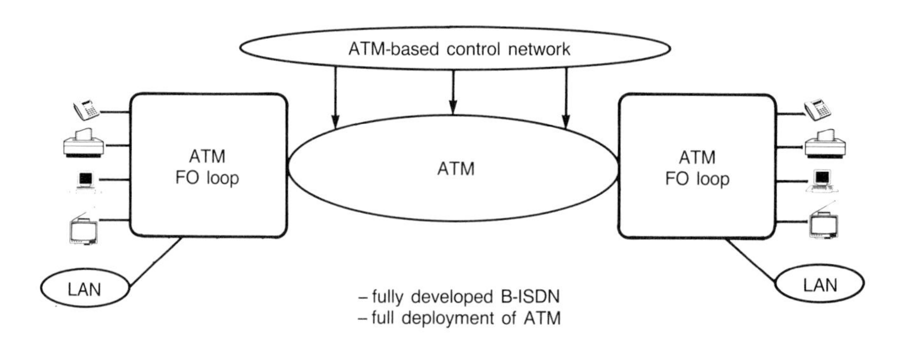

networks for the 2000 era

Fig. 1.2 Introduction of fibre-optic subscriber loop networks and ATM networks.

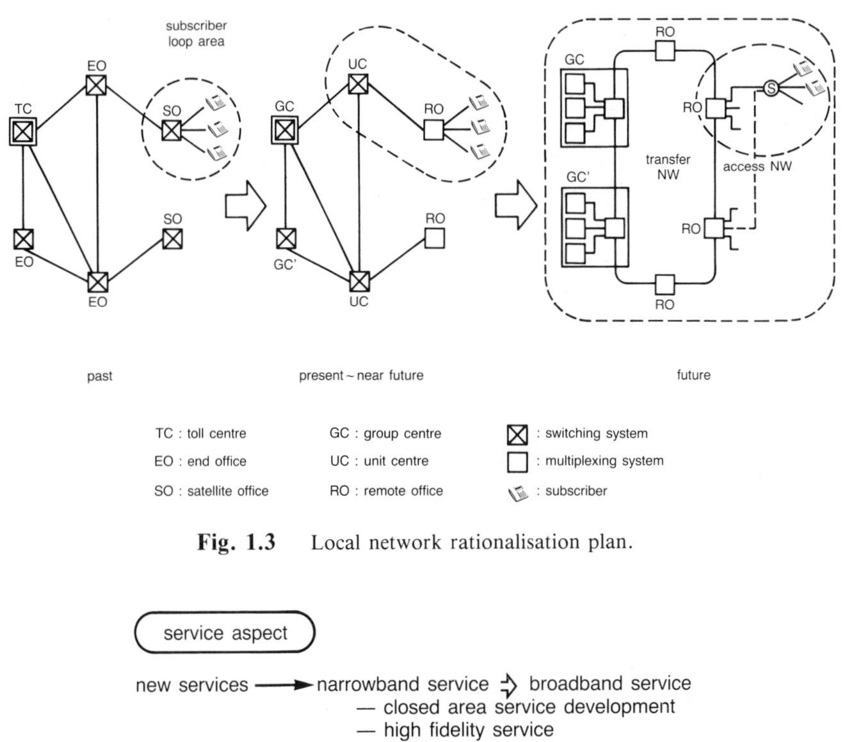

Fig. 1.3 Local network rationalisation plan.

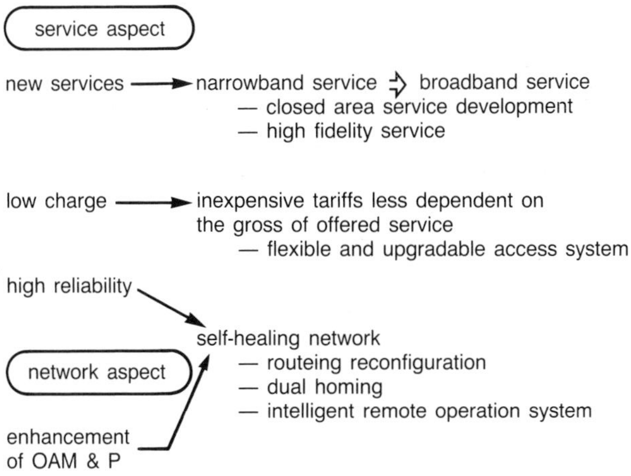

service aspect

new services → narrowband service ⇨ broadband service
— closed area service development
— high fidelity service

low charge → inexpensive tariffs less dependent on the gross of offered service
— flexible and upgradable access system

high reliability → self-healing network
— routeing reconfiguration
— dual homing
— intelligent remote operation system

network aspect

enhancement of OAM & P

Fig. 1.4 Subscriber loop network features in the future.

1.3 SERVICE ASPECTS

Before the specifications of the future fibre-optic subscriber network can be determined, its intended services must be clarified. There are two possible service scenarios — narrowband services are provided exclusively or a combination of narrowband plus video distribution services is supported. It can be generally said that narrowband services are more popular and achieve higher penetration rates. Narrowband services evolve more frequently and are more suitable for residential subscribers. Thus the fibre-optic subscriber network must be designed to accommodate existing narrowband services while stimulating the creation of new narrowband services.

The level of Japanese residential subscribers contracting more than one POTS circuit in 1989 was 7.1% and is rapidly growing, as shown in Fig. 1.5. If new narrowband services were provided, the number of subscribers requiring additional circuits would increase considerably. The broadband capacity of fibre-optic transmission makes it possible to offer various closed area services, such as hi-fi audio distribution and PC communication, as illustrated in Fig. 1.6. These closed area services could be provided at an inexpensive rate not strongly dependent on the number of services offered. The cost savings and service quality of these services will ensure rapid customer acceptance. In addition to system cost reduction, the development of new cost-effective narrowband services is the key for large-scale fibre-optic loop migration.

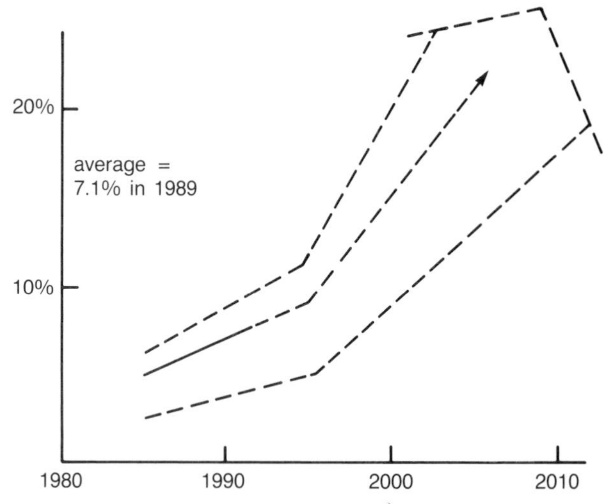

Fig.1.5 Demand for more than 1 POTS in residential use in Japan.

For large businesses, various services are already being provided through fibre-optic loop systems. From 1995, fibre-optic loop systems will be introduced to small businesses and residences, then POTS, 2B + D, low-speed data dedicated services as well as TV distribution might be provided.

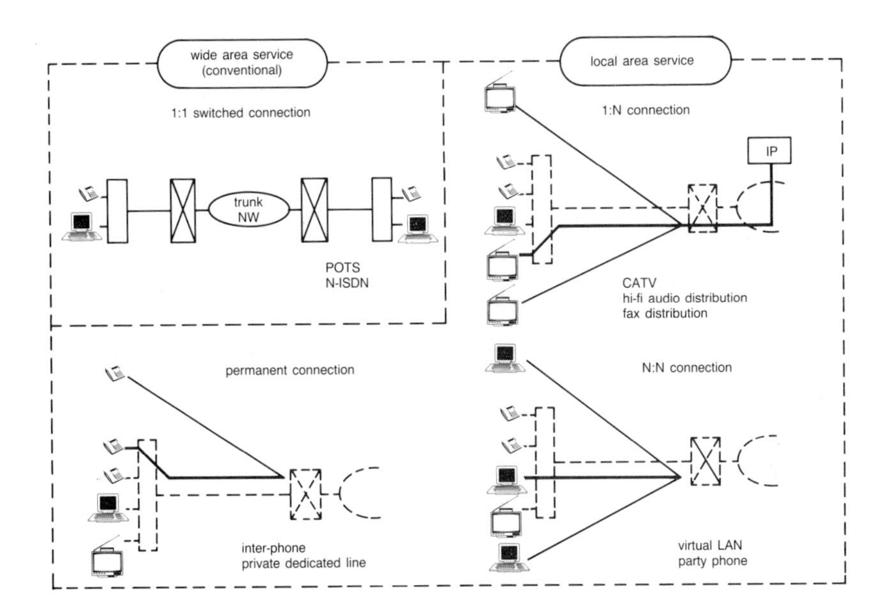

Fig. 1.6 Closed area services with the fibre-optic subscriber network.

Figure 1.7 shows one possible service evolution plan which was created by considering the demand for each service.

Even though analogue FDM transmission will be initially adopted, digital transmission will be applied later for both regular TV and HDTV signals. A recent discrete cosine transform (DCT) coding technique can be used to transport regular TV and HDTV signals with broadcasting quality at rates of 5 Mbit/s and 30 Mbit/s respectively. Figure 1.8 shows recent picture-quality evaluations for band-compression technologies. Such low bit rates make digital TV FDM transmission possible with the conventional CATV frequency band allocation. If this does not prove to be feasible, it is possible to distribute a large number of digital TV channels by TDM transmission at a reasonable bit rate.

Digital cellular radio systems of the future will utilize local loop networks between distributed antenna sites and a central office. The direct analogue transmission of radio frequency signals is a promising application for fibre-optic loops. Thus newly developed fibre-optic subscriber loop networks should be designed to cover cellular radio applications.

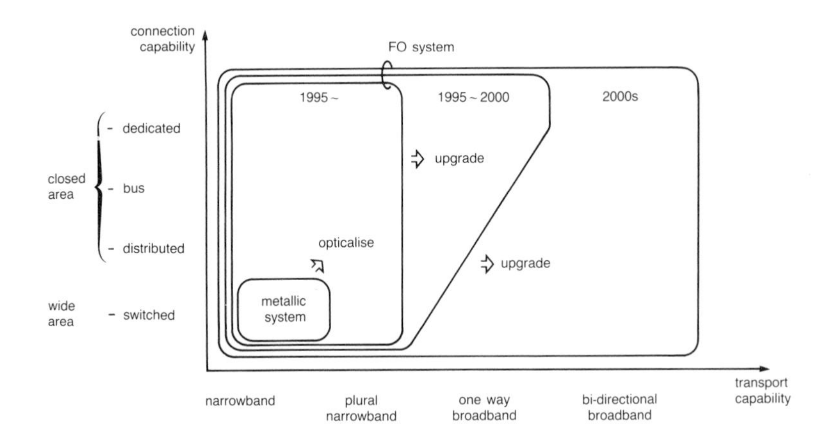

Fig. 1.7 Summary of service evolution.

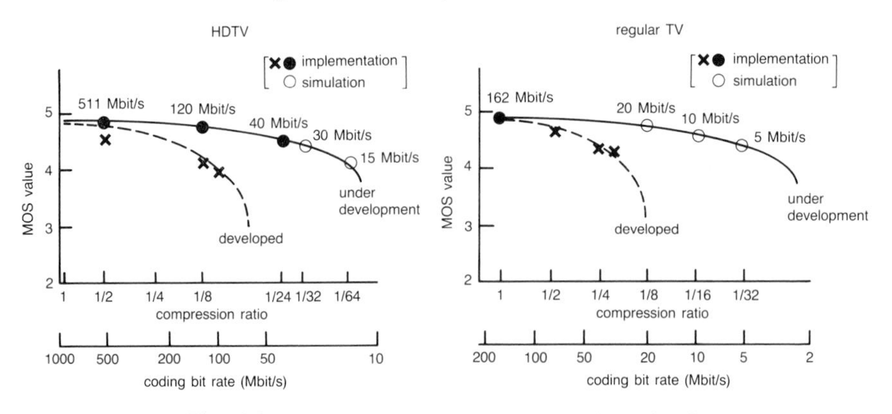

Fig. 1.8 Bandwidth compression of television signals.

1.4 INTEGRATED FIBRE-OPTIC SUBSCRIBER NETWORK

The introduction of a fibre-optic subscriber network is a golden opportunity to realize the network evolution concept as previously described.

Figure 1.9 shows the fundamental idea of the IFOS (integrated fibre-optic subscriber) network which is proposed for the future subscriber network based on this network evolution concept [5]. The proposed network consists of two layers — the transfer network and the access network. There are four main types of access system to be introduced. They are indicated by N, V, H and R in Fig. 1.9. The transfer network has a ring topology and dual homing to the central office for high reliability. The proposed access network

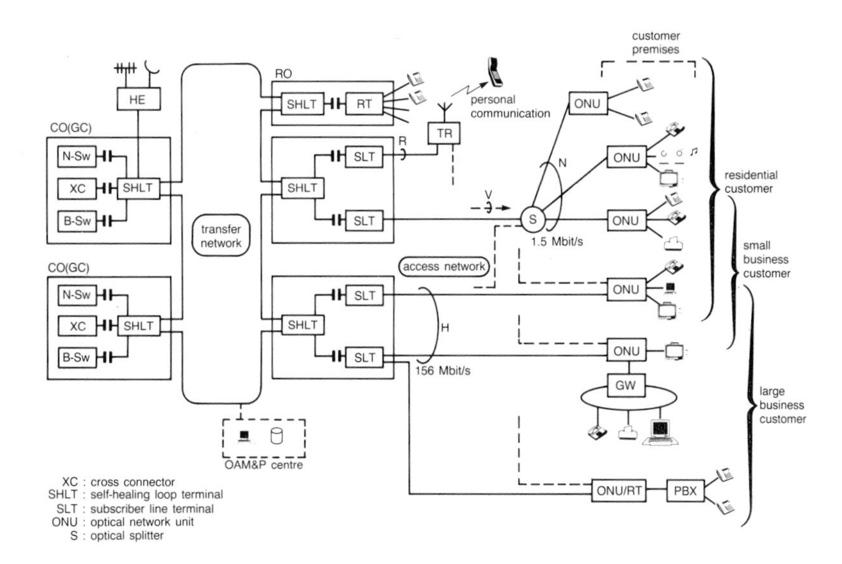

Fig. 1.9 Development target of IFOS.

uses the concept of double routeing and dual homing to remote office from each subscriber. The dotted lines of the access network in Fig. 1.9 show the dual homing routes.

System-N is a PDS (passive double star) system using optical splitters with possibly 8-16 branches. It can offer a variety of services totalling about 1.5 Mbit/s to each customer. The PDS system is most promising because it offers the lowest cost per subscriber [8,9].

System-V distributes video signals, such as dozens of regular TV channels and/or HDTV. The recent development of erbium-doped fibre amplifiers has made multichannel TV signal distribution using the AM and/or FM FDM formats practical and cost-effective [10].

System-N and System-V permit fibre sharing by employing WDMs. The combination of System-N and System-V will satisfy the near-term transport demands of residential customers as well as small businesses.

System-H is a single-star, high-speed digital system which can simultaneously provide a wide range of multiplexed transport capabilities. System-H must satisfy business customers who require substantial high-speed services.

System-R is dedicated to cellular radio systems, especially for the personal communications networks that require extremely large numbers of base antenna sites in town and along the highways.

In order to pursue FTTH, the most cost-effective design for System-N is very important as well as the provision of new services as described in section 1.3.

Figure 1.10 shows the transport capacity and PDS branches for System-N. The narrowband transport capacity required by either a residential customer or by a small business customer in the year 2000 has been roughly estimated. This estimation took into account POTS/ISDN switching circuits, hi-fi audio distribution, dedicated lines, virtual LAN channel and so on, by reference to individual penetration rates and the growth rate of existing similar services in Japan. The estimation indicates that 12 B (768 kbit/s) is sufficient for the average customer [11].

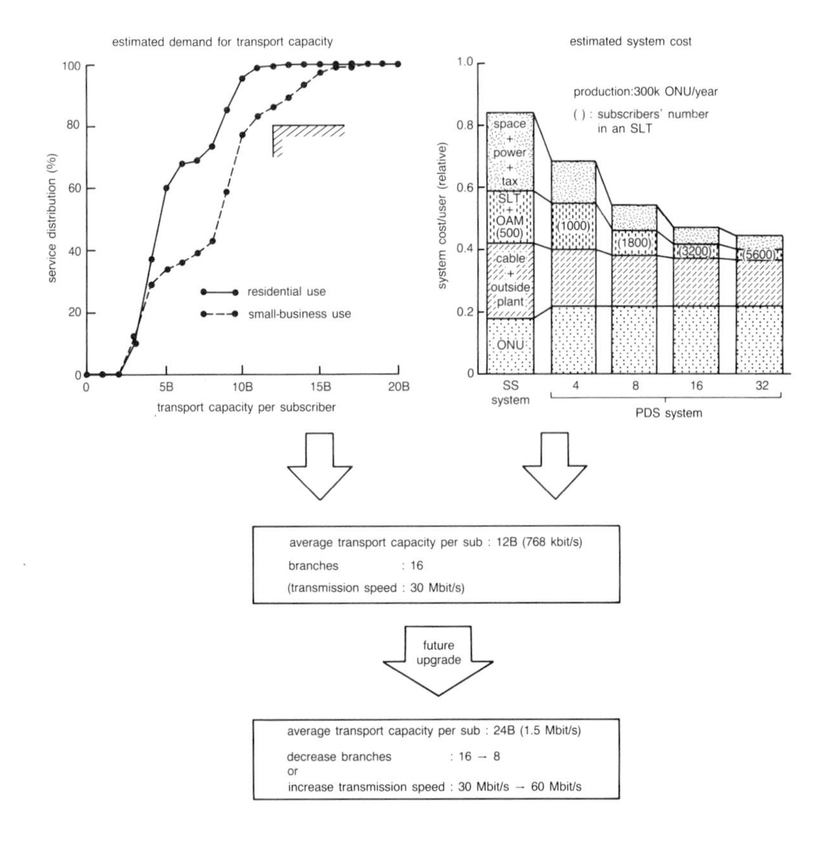

Fig. 1.10 Fundamental design of IFOS System-N for 1995-2000 time frame.

The branches of each PDS should be as small a number as possible within the range assuring the system cost-effectiveness. This is because small branching numbers give a PDS system the advantages of more flexible deployment and easier upgrades in the future. Taking into account the utilization of advanced CMOS devices, the 16-branch is most suitable for our System-N design [11]. There are two schemes to cope with increasing narrowband service demands in the future; one is to reduce branches further and the other is to increase transmission speed.

It is very important to be able to upgrade towards a bi-directional broadband transport capability. Figure 1.11 shows upgrade schemes for a system with a PDS architecture. The phase 1 system transports narrowband bi-directional 1300 nm signals at up to 1.5 Mbit/s for each customer by employing the TDMA and TCM (time compression multiplexing) techniques. Multichannel TV-FDM signals are transmitted at 1550 nm. As for video services, AM and/or FM FDM systems would be most cost-effective for the time being, considering their compatibility with existing TV sets and/or BS tuners.

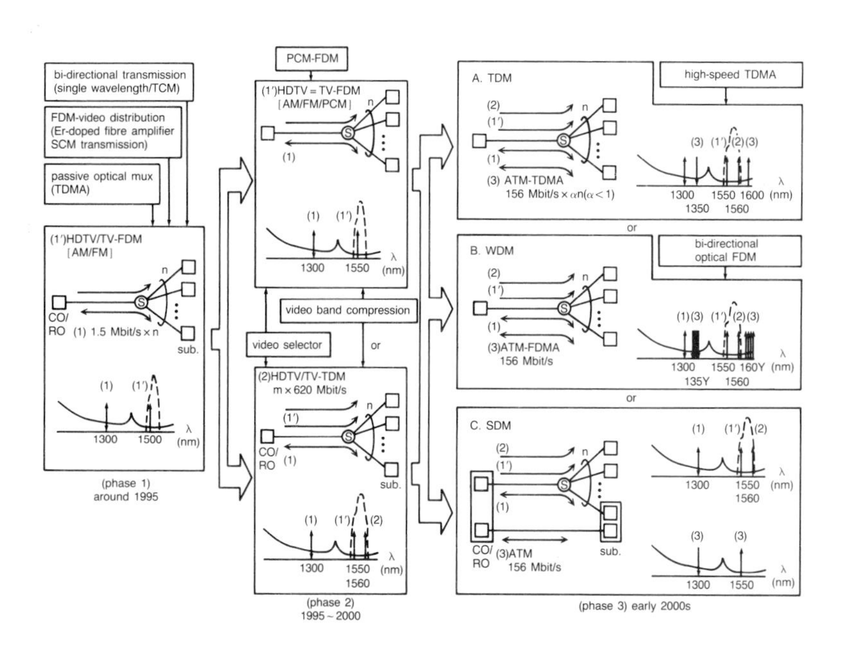

Fig. 1.11 Upgrade scheme for small business and residential application.

In phase 2, which sees the emergence of PCM-TV broadcasts, there are two upgrade schemes that are compatible with the phase 1 system. One is to carry new PCM-TV signals in spare FDM channels. The other is to add new wavelengths, for example 1560 nm, and to transmit new PCM-TV signals in the TDM format. Downstream switched video services will also be provided in phase 2.

In phase 3, there are three ways of upgrading to a fully bi-directional broadband ATM connection. They are shown in Fig. 1.11 and are the three well-known multiplexing techniques, namely TDM, WDM and SDM. In WDM, *n* subscribers' numbers are allocated for *n* different downstream wavelengths in the 1600 nm band, and *n* different upstream wavelengths are allotted in the 1350 nm band. Additional fibres are needed for SDM. Because each technology is growing rapidly, the final choice cannot be made at present.

Figure 1.12 shows a cost estimation for System-N over the 1995 to 2000 time frame, together with the conventional metallic system cost. In Fig. 1.12, the cost unit of 1 means the conventional POTS construction cost in Japanese urban areas, and production scale is 300 000 ONUs per year per vendor. The fibre-optic system cost target is roughly equal to two metallic lines of POTS or one 2B + D ISDN service when offering 2POTS or 2B + D. There are various combinations of narrowband services. Fibre-optic system cost would be less dependent on the gross of the services, in contrast to existing metallic system cost which is almost directly proportional to the gross of the services. This means that a fibre-optic subscriber network could provide various combinations of narrowband service at a charge nearly equal to the basic monthly charge of the existing 2B + D ISDN service.

1.5 SYSTEM DEPLOYMENT STRATEGY

The service development efforts and the pricing strategy described above should lead to the service penetration rates shown in Fig. 1.13. In this figure, FO (fibre-optic) service means 2B + D plus additional services which can be easily provided by a fibre-optic system.

There are two typical approaches to installing a fibre-optic subscriber loop network — to opticalize subscribers only when they demand FO services and 2B + D services (approach I), or to opticalize all subscribers in a specific area on a planning basis (approach II). These are shown in Figs. 1.14 and 1.15 respectively. A cost comparison of these approaches was performed and the results are shown in Fig. 1.16. The four main assumptions were that opticalization would be complete in 2015, demand would increase as shown in Fig. 1.13, labour costs would continue to increase and equipment costs would decrease. Furthermore, it was assumed that fibre cables in

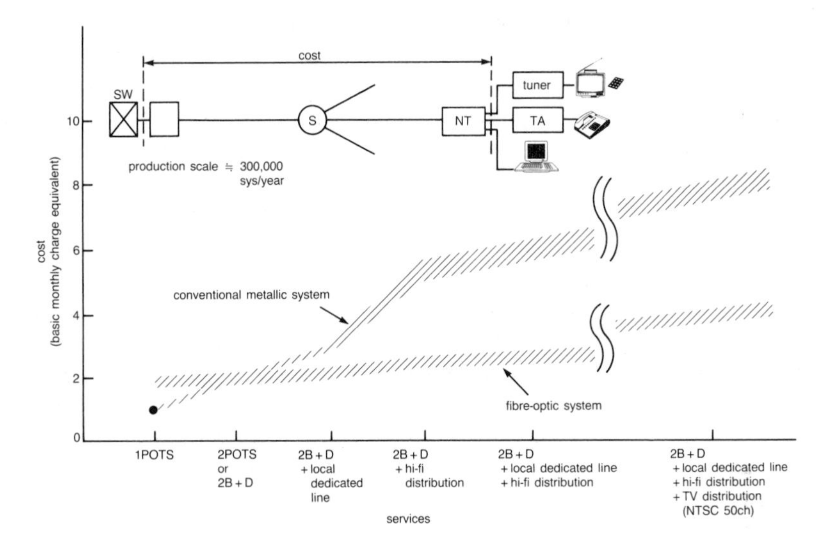

Fig. 1.12 Cost estimation for IFOS in 1995-2000 time frame.

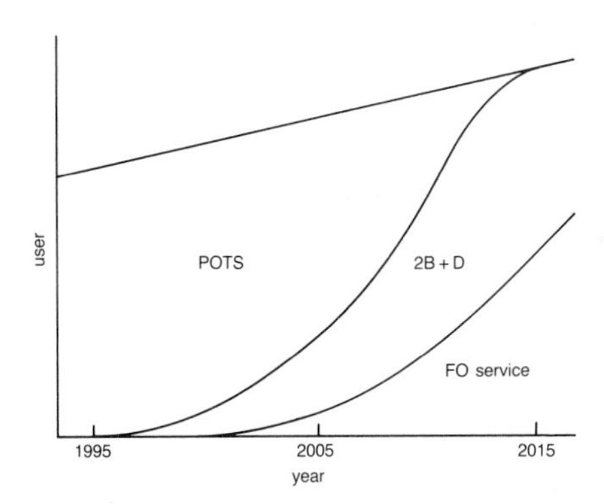

Fig. 1.13 Service development — estimated demand.

approach I always suffer from low connection rates, i.e. many optical fibres remain unused. The converse was assumed for approach II.

According to this cost comparison, over the 20-year period from 1995 to 2015, approach II requires 15% less expenditure than approach I [12]. Another factor supporting the acceptance of approach II is that it encourages the introduction of new closed area services as described in section 1.3. In the completely opticalized environment, a variety of new closed area services such as on-demand hi-fi, virtual LAN and party-phone can be provided effectively to all customers in connected areas.

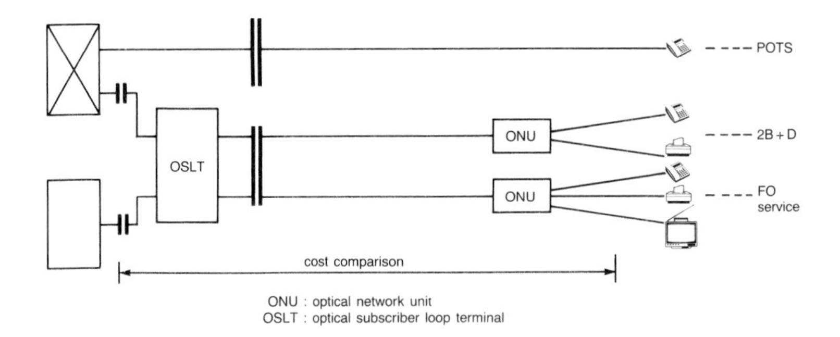

Fig. 1.14 Service development — deployment approach I.

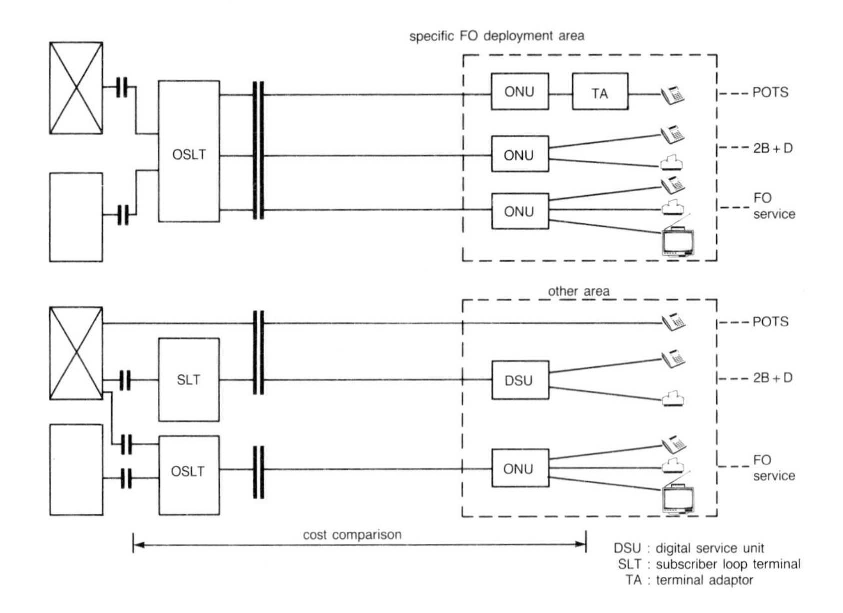

Fig. 1.15 Service development — deployment approach II.

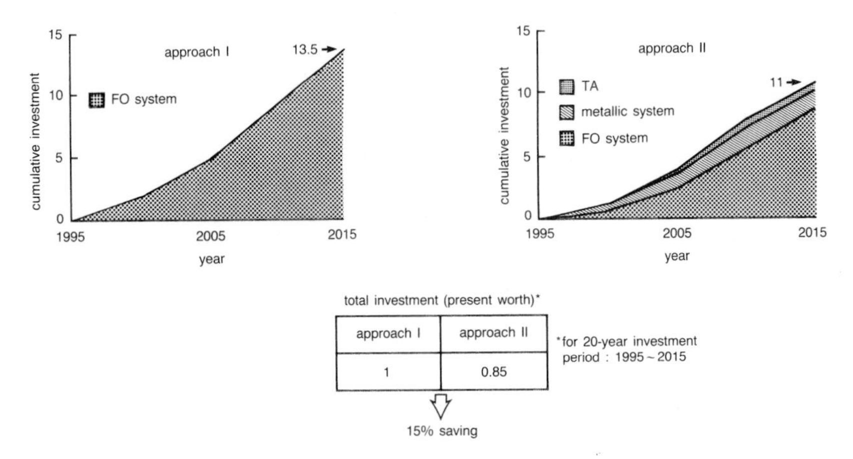

approach I	approach II
1	0.85

Fig. 1.16 Estimated investment for fibre-optic subscriber network.

1.6 CONCLUSION

The goal of NTT is to provide visual, intelligent and personal (VI&P) services to its customers [13] through a completely optical communications system by the year 2015. The key to achieving this is the development of an effective fibre-optic subscriber loop network and its efficient introduction. Effective means that the network supports popular narrowband services and encourages the development of new services. The IFOS network is proposed because it can support any combination of narrowband-visual, high-speed, and radio systems. The narrowband system will be the basic system that gains widespread acceptance from the public. Its targeted system cost is just twice that of the conventional metallic POTS, while offering 2B + D services. IFOS was shown to significantly reduce operation costs while improving network reliability.

To determine the optimum introduction strategy, a detailed cost comparison was performed. It showed that IFOS should be introduced on an area-by-area basis not subscriber by subscriber.

IFOS will form the infrastructure needed for BISDN and so realize NTT's VI&P communications vision.

REFERENCES

1. Habara K: 'ISDN: a look at future through the past', IEEE Comms Magazine, 26 , No 11, pp 25-32 (November 1988).

2. Turner J S: 'New direction in communications (or which way to the information age?)', IEEE Comms Magazine, 24 , No 10, pp 8-15 (October 1986).

3. Takashima S: 'Introduction of fibre optic subscriber networks', Japan Telecomm Review, 30 , No 4, (October 1988).

4. Boyer G R: 'A perspective on fibre in the loop systems', IEEE Magazine of LCS, 1 , No 3, pp 6-11 (August 1990).

5. Miki T: 'Introduction plan for optical network of the BISDN', 16th ECOC, 2 pp 771-778, Amsterdam (September 1990).

6. Kimura H, Niwa A and Suzuki S: 'Basic scheme for the INS network', Review of ECL, 33 , No 5, pp 751—756 (May 1985).

7. Miki T, Kano S, Inoue Y and Yamaguchi H: 'Lightwave-based intelligent transport network', ISSLS '86, p 47, Tokyo (September 1986).

8. Tokura N, Oguchi K and Kanada T: 'A broadband subscriber network using optical star coupler', IEEE Globecom87, 37.1, Tokyo, pp 1439-1443 (November 1987).

9. Stern J R et al: 'TPON - a passive optical network for telephony', 14th ECOC, pp 203-206, Brighton (September 1988).

10. Kikushima K and Yoneda E: '6-stage cascade erbium doped fibre amplifier for analogue AM- and FM-FDM video distribution systems', OFC'90, PD22, San Francisco (February 1990).

11. Shinohara H and Yamashita I: 'Service evolution and deployment scenario for IFOS systems', Workshop on Passive Optical Networks, 2.4, San Diego (December 1990).

12. Okada K, Shinohara H and Miki T: 'Introduction approach toward fully fibre optic access network', ICC'91, Denver (June 1991).

13. Yamaguchi H: 'A view toward telecom services and networks in the 21st century', NTT Review, 2 , No 2, pp 6-17 (March 1990).

2

OPTICAL FIBRE PILOT PROJECTS

W Rosenau

2.1 INTRODUCTION

Deutsche Bundespost (DBP) recognized the far-reaching significance of optical fibre technology at a very early stage in terms both of its technical/commercial and of its economic aspects and has made great efforts to use this modern technology as the main component of a highly efficient telecommunications infrastructure.

Owing to the law concerning the structure of posts and telecommunications of 1 July 1989 and Deutsche Bundespost's division into three enterprises, DBP Telekom (Telekom) has been working as a public enterprise since the beginning of 1990 [1].

Telekom operates within a stable regulatory framework which also applies to optical fibre activities.

This stable regulatory basis, which basically entails the maintenance of the monopoly for installing and operating all transmission paths (cable, transmission systems, microwave) and for the telephone service, allows Telekom to continue its optical fibre introduction strategy developed by the telecommunications branch of the Deutsche Bundespost and to integrate current requirements.

The strategy for introducing the optical fibre technology pursued so far in western Germany has been marked from the very beginning by two objectives:

- to implement an economically optimized network design for established telecommunications services by using this technology as a standard system at the long-distance level and for the interconnection of the new digital exchanges at the local level;
- to provide an initially limited optical infrastructure (overlay network) in the vicinity of the subscriber, thus creating the conditions for the development and testing of new broadband services and meeting the initial demand in this segment.

On the basis of this double-track introduction strategy Telekom laid about 1 million optical fibre kilometres in its telecommunications network by the end of 1990, 768 000 fibre kilometres of which were attributed to the long-distance network and 262 000 to the local network.

In the local network about 25% of the fibre kilometres in the access line network were used for developing the overlay networks for new broadband services.

The new strategic approach of Telekom, called 'fibre to the home' (FTTH), aims to establish the desired optical fibre infrastructure through the commercial use of established services (telephone, data transmission/communication, radio and TV distribution, etc), and also to keep open the option of upgrading the new optical fibre systems as and when required. In addition the provision of low-cost broadband communication forms will create a good basis for the further development and acceptance of these services and the incorporation of Telekom's expertise in this field.

Owing to the technical progress, this new approach leads to the return from the double-track introduction fibre strategy to a single-track strategy. It also marks corporate restructuring by extending the (standard) use of optical fibres to all network levels. As to the local network, the optical fibre's ability to offer new broadband services is no longer understood as a prerequisite but as an 'option for the future' [1].

The FTTH activities of Telekom have gained a new dimension owing to the unification of Germany. This provides a chance to use optical fibre to improve the poorly developed telecommunications infrastructure in eastern Germany as quickly and completely as possible.

With its activities concerning FTTH or fibre in the loop (FITL), Telekom is pursuing a pragmatic solution which is designed to bring about quick progress. A major element of this solution is the fast implementation and evaluation of a series of pilot projects called 'optical access line' (OPAL). This series of pilot projects is presented in this chapter.

The following explanations concerning the state of the art of the telecommunications networks in eastern and western Germany and concerning the early testing of optical fibre at the subscriber line level reflect once again the reasons why Telekom is developing and providing FITL systems in its sphere of responsibility.

2.2 PRESENT SITUATION IN THE TELECOMMUNICATIONS NETWORKS

As to the use of optical fibres at the subscriber line level in the former Federal Republic of Germany, the following areas are of top priority:

- the telephone network;
- the integrated data and text network;
- the broadband distribution network.

The degree to which cable-based networks have been established in western and eastern Germany differs greatly [1].

In western Germany the situation is as follows.

- Telephone network — the hierarchical telephone network is a well-developed nationwide network. At the end of 1991 some 31.3 million subscribers were connected to 6200 local exchanges in 3500 local networks (Fig. 2.1); by the end of the year 2000 the subscriber lines will increase to some 40 million.
  The telephone network will continue to be digitized. By the end of 1993 about 30% of the existing access capacity will be serviced over digital local exchanges.
  The access cable network of a local exchange is star-shaped. The feeder cables beginning at the exchange lead to a distribution unit, the so-called cable distribution box, from which distribution cables lead to the network terminations in the serviced buildings. The feeder and distribution cables have an average length of 1700 m and 300 m respectively.
  About 90% of the feeder and distribution cables have been laid as conduits or ground cables; the number of overhead lines is low (10%) and is constantly falling.
  At present, about 13 000 analogue front-end units have been installed in front of local exchanges; however, digital front-end units have been installed only to a limited extent so far.
- Integrated data and text network — some 600 000 data stations have been connected to the nationwide integrated data and text network (IDN) that was established in 1978 by interconnecting several special networks for economic reasons.

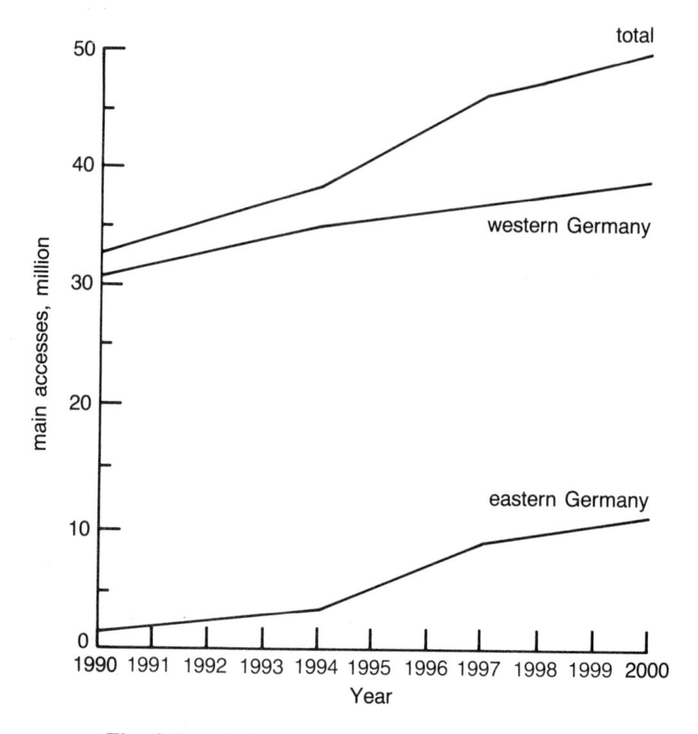

Fig. 2.1 Main accesses (development 1991—2000).

IDN has two switching systems that enable the use of a large service spectrum (Fig. 2.2) over circuit and packet-switched connections. The physical cable network of IDN is founded on the basic infrastructure of the telephone network. The subscriber lines are established in the local network on the basis of the telephone network's pool of lines; here specially selected lines may be used.

- Broadband distribution network for radio and TV broadcasting — since the end of the 1970s Telekom has established and operated coaxial cable distribution networks at the local level to provide households with radio and TV broadcasting programmes.
  During the past nine years the increasing number of programmes and the excellent transmission quality have resulted in a rapidly increasing demand for access, and a growing number of local broadband distribution networks. At the end of 1991 already 9.79 million of the 26.3 million households in western Germany were connected to the broadband distribution networks (Fig. 2.3) and for another 7.81 million

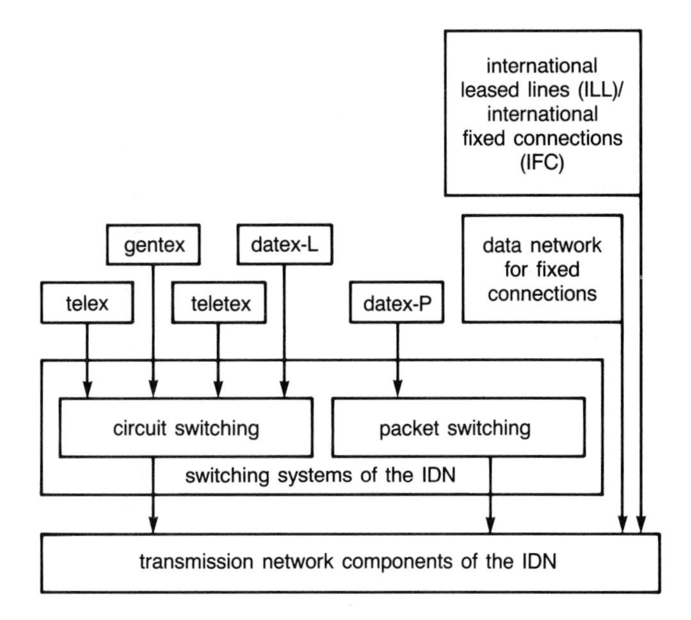

Fig. 2.2 Integrated text and data network (IDN) — services provided.

households the technical prerequisites for connection to the existing distribution network have been met.
The tree-shaped broadband distribution networks transmit a frequency band of up to 450 MHz, thus offering up to 38 AM/TV programmes as well as 16 digital- and 30 analogue-modulated radio programmes.

The present situation in eastern Germany, in comparison, is much less favourable.

- Telephone network — the telephone network's state of the art is unsatisfactory.
  At the end of 1991 about 2.4 million telephone accesses were connected to 2674 local exchanges in 1474 local networks, the local subscriber lines being marked to a large extent by multiple use. With its project 'Telekom 2000' Telekom will increase the number of telephone accesses to some 9 million by the year 2000 and at the same time practically replace the existing access cable.
  The necessary rebuilding of the local networks in the east of the country by installing only digital local exchanges will lead to considerable growth rates per year in the cable networks. Owing to the lack of reserves the growth rates will be much higher than in western Germany.

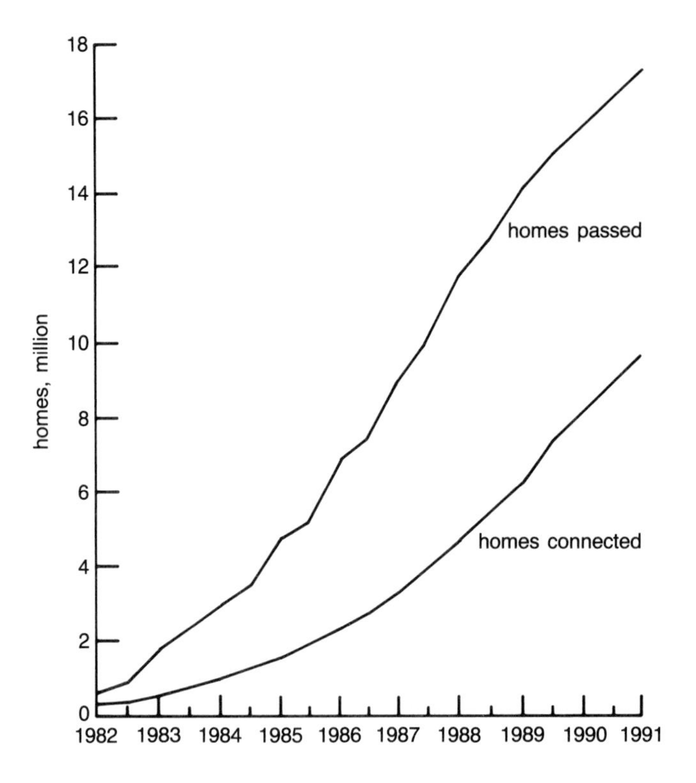

Fig. 2.3 Broadband distribution network — development of cable access in western Germany.

- Integrated data and text network — the IDN is currently being established. Its network structure still differs from the one existing in western Germany; as it is not yet possible to determine where demand will be concentrated in the future, the 14 network nodes that have been set up cover a wider area than is the case in western Germany.
  Broadband distribution networks for radio and TV broadcasting — following unification, broadband distribution networks were established in the eastern part of Germany along the same lines as in the western part of the country. At present several hundred thousand homes can be connected to the networks that have already been installed.
  The ambitious target of Telekom to upgrade eastern Germany's telecommunications infrastructure to reach western Germany's standard by 1997 is an opportunity to use FITL systems on a large scale to build up a totally new infrastructure without having to compromise with existing networks.
  In view of these prospects the development and provision of FITL systems is gaining a significance which reaches far beyond a 'technical challenge'.

2.3 OPTICAL FIBRE TESTING PROJECTS OF TELEKOM 1978-1990

As early as 1975 it became apparent that the technological developments in the field of optical transmission systems would succeed. As a result Telekom decided to carry out a series of optical fibre tests for gaining experience and insights from the practical use of various research, trial and operational systems [2].

From the very beginning the testing projects were based on two developments resulting in:

- the fast provision of multichannel and digital optical fibre transmission systems to be used in the long-distance network and the junction line network of the local networks;
- the early use of optical fibre technology at the subscriber line level for providing narrowband services and new broadband communications modes.

The optical fibre test projects Berlin I, II and III aim to test optical fibre itself, i.e. its behaviour under operational conditions. These projects mark phase one of the optical fibre subscriber line tests.

However, another big step was necessary to actually implement the theoretical prospects — the transition from the technology development to the application-related overall system or — expressed in a different way — from the optical fibre transmission technology to a services integrated local access technology. This step was the system trial BIGFON (Breitbandiges Integriertes Glasfaser-Fernmeldeortsnetz — broadband integrated optical fibre local telecommunications network).

The BIGFON system trial which was closely observed worldwide aimed from the outset at exploiting the industry's entire expertise in a kind of competition of concepts with the objective of using optical fibre technology at the subscriber line level and exploiting a future-oriented concept to provide subscribers with all presently available and a range of future narrowband and broadband services for both individual and mass communications.

What was essential for the BIGFON system trial was that Telekom confined itself to some crucial overall requirements to give industry the chance to develop widely varying concepts in a competition of ideas. A wide range of companies and consortia participated in the system trial which to a great extent was paid for by Telekom. It was clear from the beginning that on account of numerous unspecified technical conditions and the lack of international standards, there could be no competitive bidding procedure on the introduction of standard equipment at the end of the trial. However, Telekom and industry gained important knowledge about the problems still to be solved before FITL systems can be introduced on a large scale.

One of the most important insights gained from the BIGFON system trial, which began in 1983 and ended officially in 1986, was the fact that, in spite of the fibre's almost unlimited bandwidth, the simultaneous transmission of several broadband signals (e.g. several TV programmes) to the subscriber's home was limited for technical reasons. Therefore, alternative techniques to provide the subscriber with a wide range of programmes had to be tested, such as the selection of the TV programmes by the subscriber (video on demand) in the exchange. Despite the great progress made with fibre and components, similar considerations have to be made even today, particularly in view of the introduction of HDTV.

For the reasons already mentioned and because low-cost, optical and electrooptical components could not be expected to be available within a short period, the BIGFON system trial could not provide an optical fibre access system to be directly implemented as a large-scale system for the provision of narrowband and broadband services.

However, the results of the system trial stimulated an intensive and partially public discussion leading to a new approach to introducing optical fibre technology at the subscriber level.

This approach envisaged the limited development of a minimum optical fibre infrastructure reaching to the vicinity of the subscriber in selected local networks which would provide a preparatory network for new interactive broadband services.

The implementation and consistent development of this approach resulted in:

- the development of optical fibre overlay networks in 29 large western German cities and the western part of Berlin [2];
- the establishment and operation of the digital and subscriber dialling switched broadband network (VBN) — on the basis of overlay networks — which has offered interested customers the opportunity to experiment with new broadband communications modes and to gain know-how since the beginning of 1989 [2];
- the introduction of the associated Berlin Kommunikation (BERKOM) project to develop and test
  - — transmission systems (e.g. ATM),
  - — applications,
  - — terminal equipment,

for the broadband area and on this basis to elaborate proposals for the standardization of broadband services.

The optical fibre pilot projects OPAL presented below are directly related to the BIGFON system trial. They take up the old tasks and are seeking an economically acceptable solution by using advanced technical possibilities.

The large-scale optical fibre systems derived from the OPAL pilot projects provide new prospects for broadband activities (overlay network, VBN, BERKOM) and for making use of insights and experience gained in the broadband area.

2.4 USE OF OPTICAL FIBRE TECHNOLOGY AT THE ACCESS LINE LEVEL

2.4.1 General conditions

In Telekom's sphere of responsibility the use of optical fibre systems is marked by numerous, sometimes contrary, general conditions that can be listed here only to a certain degree.

- Telekom operates networks for individual and mass communication and at the same time has a monopoly on the installation and operation of all transmission paths. This situation means:
  - no restrictions on the topologies and architectures to provide a low-cost infrastructure in the local network that integrates the networks and services of Telekom by exploiting the efficiency of the optical fibre system;
  - the obligation to design an 'open' optical fibre infrastructure that can be used at acceptable cost and without restrictions by other service providers (open network provision (ONP)).
- Telekom has an infrastructural obligation derived from the monopoly on the installation and operation of all transmission paths and the telephone service (live speech). Particularly in view of establishing FITL systems in eastern Germany this infrastructural obligation requires action because the large-scale use of copper cables is avoided if FITL systems are implemented early. At the same time this allows the chance to use totally new and optimized infrastructures. However, it is essential for Telekom to co-ordinate its FITL activities with the activities of network operators in other countries in order to concentrate demand for optical

fibre access systems and thus to reduce costs of optical and electrooptical components.

- A variety of systems is bound to be established as Telekom is confronted with the necessity to introduce optical fibre systems to provide regions marked by totally different degrees of development (eastern and western Germany) and differently structured customer segments (private and business customers) with different service combinations. However, this diversity of systems should be reduced as much as possible for operational and technical/economic/organizational reasons.

The general conditions outlined here, and others not mentioned, have a great influence on the specification of the optical fibre structures and the large-scale FITL systems in Telekom networks. At the same time they impede co-ordinated activities with network operators in other countries with a totally different legal and regulatory environment which leads to different implementation plans for FITL systems.

2.4.2 Requirements to be met by optical fibre systems

Telekom formulated the following general requirement profile for the development of optical fibre systems:

- costs equivalent to costs of comparable copper systems if purchasing quantity is the same;
- compatibility with existing or planned transmission and switching systems of Telekom;
- option of including future technological progress, particularly in the feeding-in and feeding-out of the signal and the use of amplifiers;
- network structures providing a greater safety factor against faults and cable interruptions;
- development of a transition technology to link the new optical fibre networks with the existing copper networks;
- possibility of integrating the optical fibre networks into Telekom's overall network management concept with stand-alone solutions being allowed on a transitory basis;
- option to extend the service offerings (broadband services) owing to cost-effective further development of systems.

These technology-oriented requirements must be gradually elaborated in more detail taking into account the general conditions for the use of optical fibre technology in section 2.4.1.

2.5 PILOT PROJECTS

2.5.1 Targets

The targets of the optical fibre pilot projects (OPAL) of Telekom can be structured in the following way.

The global target is the testing of different innovative concepts leading directly to a system decision and thus to the early provision of (not yet fully standardized) fibre systems that:

- are cost-effective;
- can be integrated;
- can be optionally upgraded to offer new services.

These systems are to ensure **today's** provision of diverse customer segments based on the available services suitable for combination.

The global target encompasses a second detailed target that, via a comparison of tested and viable concepts and the localization, description and preparation of detected and expected problem areas, leads to the development and specification of facilities for different fibre systems and the underlying optical fibre network topologies and thus to the determination and description of the necessary specifications and standardizations.

This way the production and introduction of a second generation of standardized fibre systems is prepared and the ability of Telekom to evaluate the purchase of these systems on the world market is ensured.

The third target level encompasses the testing of the service-related, new technical and operational possibilities of optical fibre technology which are presently of secondary importance, for example:

- the opening up of new transmission capacity (e.g. extension of the transmitted frequency band in the broadband distribution network);
- the modification of today's services;
- the future-oriented concept of operation control and support systems.

2.5.2 Selection of pilot systems

Telekom has been searching for technical concepts — originally meant for western Germany — which on the one hand enable the establishing of the desired FITL infrastructure through the commercial use of established services (telephone, data transmission/communication, radio and TV distribution, etc) and on the other hand enable the upgrading of these new optical fibre systems as and when required for a low-cost provision of broadband communication forms. This search has led to two activities.

The first activity was the European-wide competition of concepts 'Economic use of optical fibre technology at the subscriber line level' initiated in mid-1989. Sixteen companies and consortia submitted alternative concept proposals thus showing industry's vivid interest.

The competition of concepts was:

- to enable Telekom to select, from among the new concepts for innovative optical fibre systems, those concepts that are suitable for its needs;
- to test those future-oriented concepts which were considered suitable for, and worth implementing in, OPAL pilot projects.

The requirements laid down in the request for proposals were kept to a minimum to achieve the desired overview on existing concepts.

The optical fibre system concept proposals submitted to cover the services/service combinations shown in Table 2.1 were evaluated by the Telecommunications Engineering Centre of Telekom by mid-1990.

Table 2.1 Competition of concepts — fields of application.

a) CATV — in general — in rural areas — at junction line level	b) Fixed connections/switched services for analogue and digital (up to 2 Mbit/s) interfaces — in general — for business accesses in cities	c) Combinations of CATV and fixed connections/ switched services

The evaluation resulted in the decision:

- to implement four pilot projects (OPAL 4-7);
- to launch simultaneously initial procurement measures for such systems that can be considered sufficiently tested and thus ready to go into production. The procurement measures covered the optical fibre systems that — compared with the established copper systems — can be used cost-effectively at the junction line level of the broadband distribution network and at the feeder cable level of the telephone network/IDN.

Parallel to the competition of concepts the second activity was the conclusion of a co-operation agreement with Raynet Corporation, Menlo Park (USA) in July 1988.

Within the framework of this co-operation the system, originally developed by Raynet for the US market, was to be adjusted to the requirements of Telekom and further developed for low-cost implementation.

The co-operation agreement concluded with Raynet Corporation envisages two pilot projects (OPAL 1 and 2) with different tasks.

As a result of this co-operation another pilot project (OPAL 3) was initiated at the beginning of 1990 using the Raynet system technology.

The selection of pilot systems for the OPAL projects 1-7 shows the present network and service-related preferences of Telekom in relation to the use of FITL systems.

2.5.2.1 Pilot systems OPAL 4-6

The system topology and architecture was of primary importance in the selection of these pilot systems.

The pilot systems are classified as passive optical networks (PON) belonging to the subgroup of the physical star/logical bus (or passive double-star) systems.

The PON architecture, generated from the well-known (and expensive) star-shaped network by means of electronics and fibre sharing, appears quite manageable and future-proof from today's point of view and is potentially cost-effective.

The testing of these PON systems in the fibre-to-the-home (FTTH) or fibre-to-the-curb (FTTC) version shows Telekom's interest in systems that appear particularly suitable for developing a new infrastructure in eastern Germany.

The fact that six companies are involved in the pilot projects OPAL 4-6 emphasises the intention of Telekom to analyse fully and quickly the possible diversity of architectures created by the variety of company know-how.

2.5.2.2 Pilot systems OPAL 1 and 2

The selection of these pilot systems focusing exclusively on western Germany was marked by the intention to provide specific customer segments (OPAL 1: private customers; OPAL 2: business customers) with specific services (OPAL 1: telephone and broadband distribution service; OPAL 2: ISDN basic and primary rate accesses) via particularly cost-effective optical fibre systems.

The bus topology derived from the data sector and developed by Raynet Corporation seemed especially suitable for this purpose, although today this topology is viewed less optimistically in its universal applicability than a few years ago.

It is considered that this topology/architecture may be preferable for distribution services because of the considerable cost benefits expected. However, only a limited market segment for switched services is expected to be recovered.

2.5.2.3 Pilot systems OPAL 3 and 7

When selecting these pilot systems, the envisaged cost-effective provision of a specific space segment (i.e. the rural area) with the broadband distribution service was focused on.

2.5.3 Concept and implementation of the pilot projects

2.5.3.1 OPAL 1

The project agreed with Raynet Corporation in 1988 and put into operation at the end of May 1990 marks the beginning of the OPAL pilot project series.

Within the framework of the pilot project OPAL 1, up to 192 private users are provided with the broadband distribution service and analogue telephone lines.

The pilot system is based on a bus topology. The heart of the bus topology is a non-invasive coupler which supports consistent fibre and electronics sharing and thus makes possible a potentially very cost-effective implementation of a large-scale FTTC system.

The implemented pilot system combines a subsystem for the broadband distribution service with a telephone subsystem, thus further reducing the costs by combining the two underlying bus topologies in one optical fibre cable and the transmission equipment required for both subsystems on the subscriber side in one subscriber interface unit (SIU).

The private subscribers are connected to the existing 24 SIUs in the pilot area via copper cable. Each SIU can service a building with eight telephone accesses and four connection points for the broadband distribution service.

The power supply of the SIUs in the pilot project is effected centrally through a copper cable laid parallel to the optical fibre cable.

The pilot system encompasses a specific system administration module (SAM) in a stand-alone version in which the telephone subsystem can be maintained and monitored and the telephone accesses activated/deactivated centrally.

The pilot system was developed, established and installed in a very short period and with remarkably good results.

2.5.3.2 OPAL 2

In 1988 the pilot project OPAL 2 and its substance, scope and implementation duration were specified and agreed with Raynet Corporation. Here, the Raynet system technology is to be used for providing business users with:

- ISDN basic access and ISDN primary rate access;
- analogue telephone access (plain old telephone service - POTS) and access for PBXs with direct dialling.

The pilot system to be implemented in a testing area in the centre of the city (banking district) of Frankfurt/Main was implemented as a split passive optical network which deviates from the original concept.

The changing of the concept, which originally envisaged the implementation of the bus topology proven in the project OPAL 1, marked on the one hand the further development of the system technology that takes account of the conditions in the selected testing area and opens up the prospect of a widely usable large-scale system that meets even complex requirements. On the other hand it revealed apparent restrictions in the use of the bus topology which until then had been considered universally applicable.

The FTTH pilot system has three (split) supply lines with a capacity of about 200 DSOs (64 kbit/s channels) each.

The supply lines begin at an office interface unit (OIU) — with a capacity of 480 DSOs — to which the existing ISDN local exchange is connected and they reach into various directions in the pilot area.

The channel and circuit capacity of the supply lines is controlled via a system administration module (SAM). This system also enables the remote controlled maintenance and monitoring of the overall system — just like in the project OPAL 1.

The power for the optical network terminations (ONUs) on the subscriber side is supplied locally and buffered by battery. Figure 2.4 shows an installed ONU.

The pilot system was put into operation in April 1992.

2.5.3.3 OPAL 3 and 7

Today over 60% of a total of 8600 communities have access to the cable-based broadband distribution networks of Telekom in western Germany.

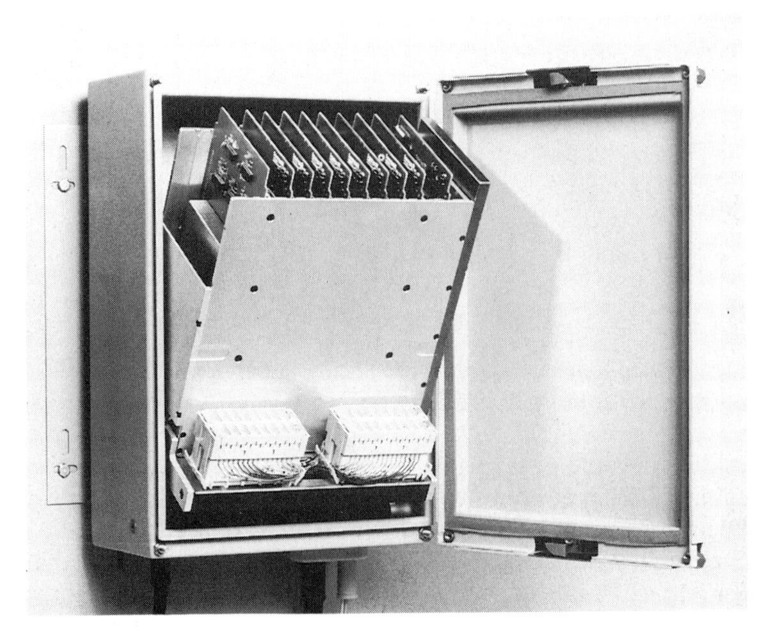

Fig. 2.4 An installed ONU.

Mainly rural areas are not provided with these networks because the use of the copper coaxial distribution system is unprofitable. For this reason the projects OPAL 3 and 7 were planned and implemented.

The pilot projects are to promote the development of optical fibre distribution systems for servicing sparsely populated areas, prove the cost-effective implementation of these systems in the trial operation and thus prepare the tapping of subscriber potential in rural areas.

The companies Raynet Corporation (OPAL 3 in Lippetal) and Bosch Telecom (OPAL 7 in Hagen near Bremen) entrusted with the provision of the pilot systems in the testing areas have independent and to some extent opposite solutions for handling the problem.

2.5.3.4 OPAL 4, 5 and 6

These OPAL pilot projects use the common passive optical networks of the type 'physical star/logical bus' for the combined provision of switched telecommunications services/fixed connections and broadband distribution services.

The pilot projects have the following features.

- OPAL 4
  The project OPAL 4 will be carried out in the Leipzig exhibition centre in Saxony.
  The pilot area includes two areas in the centre, 700 m or 1400 m from the local exchange where the central units of the pilot system are located. In each of these areas — as connecting points of the planned passive optical networks — there are six distant units (DU) for providing businesses (area 1) and households (area 2) with analogue telephone access and the broadband distribution service. The pilot system is made up of two parallel subsystems (for the telephone and broadband distribution service) based on different split optical fibre network topologies. The separate topologies both end at the DUs that are mainly located in the buildings to be serviced. Some 100 private and some 100 business pilot project subscribers are connected to the locally powered DUs via copper cable. At present, the telephone subsystem has a capacity of $12 \times 24 = 288$ channels. The transmission area of the subsystem for the broadband distribution service corresponds to the 450 MHz coaxial cable distribution system. More details are given in Table 2.2. The tender group Siemens AG, Munich, and Kathrein, Rosenheim, has been charged with carrying out the pilot project. The pilot project deals not only with providing the actual pilot system but also with building up the broadband distribution infrastructure which previously did not exist in this area (i.e. radio reception point, broadband communications amplifier station, junction line to the pilot system, etc).

- OPAL 5
  The Stuttgart-Vaihingen pilot project is a solution for combining both narrowband dialogue services and the broadband distribution service (distribution of radio and TV broadcasting) in the same network. The system contains two PON structures.

  — The optical fibre structure used for dialogue services is a passive double star with a 1:6 splitter. Customers with access to the network are provided with POTS, ISDN access and data connections via the connected ONUs; five of the ONUs are located in the homes being served and one is located in the cable distribution box.

  — The PON architecture for the broadband distribution service transmits two transmission ranges (450 MHz and 860 MHz) at two wavelengths right to the cable distribution box, where the 450 MHz band is coupled out, converted optoelectronically and routed to the connection points in the homes that are served via a passive coaxial copper cable. The 860 MHz band is routed back to the switching centre via optical fibre, where it is amplified optically and routed via a splitter to the optical connection points in the three homes that have access to the network.

Figure 2.5 shows the pilot system in schematic form.

The pilot project, which was launched in April 1992, was implemented by SELAlcatel.

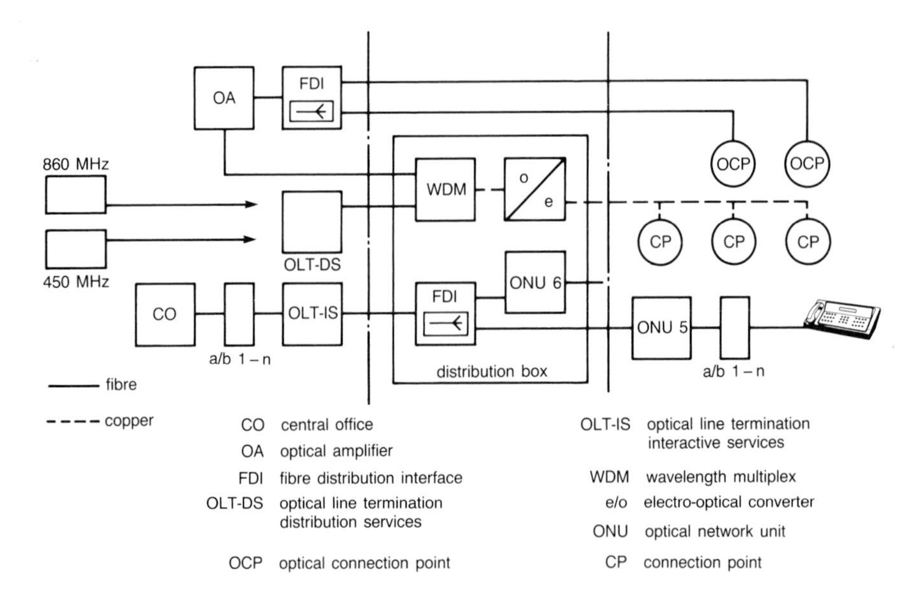

Fig. 2.5 System survey — pilot project OPAL 5 in Stuttgart.

- OPAL 6
  For the pilot project OPAL 6 a largely residential area in Nürnberg, with mainly detached or semi-detached houses and some apartment houses with up to ten apartments, was chosen. The installed pilot system provides some 200 private project subscribers with a broadband distribution service and analogue telephone access. For testing purposes one telex access, one ISDN basic access and one Datex-P access each were set up. The companies, AEG Kabel, AG, ANT Nachrichtentechnik GmbH and Philips Kommunikations Industrie (PKI) AG, formed a group which was responsible for the provision. In this pilot system the PON for interactive services begins at the central system units in the Boxdorf exchange and — as a star/bus system — ends at the subscriber side either in a subscriber

access unit (TAE), or in the vicinity of a group of small houses in a subscriber group access unit (TGAE). The TAEs, which provide one or two analogue telephone accesses each, and the TGAEs, which are installed in the houses, are powered locally by the power supply of the house of the respective subscriber and buffered by an installed battery. The signal of the broadband distribution service is transmitted via the same and a second PON:

— about 20 subscribers in a new residential area are provided with the signal up to the TAE or to an optical connection point in the house;

— however, the overall majority of the pilot system users receive the signal coupled out of the second PON and O/E converted at the two optical repeaters via a normal passive copper coaxial distribution network.

In addition it should be mentioned that the pilot area uses a new cable system that will probably help reduce the installation and, possibly, civil engineering costs.

Table 2.2 Characteristic data of OPAL pilot projects.

OPAL no.	*Location*	*Operation starts*	*Number of subscribers*	*Exchange*	*System supplier*	*Topology*	*Remarks*
1	Cologne	June 90	192	Electro-mechnical	Raynet Corp	Bus	Operating
2	Frankfurt/ Main	April 92	Approx 50	Siemens EWSD	Raynet Corp	Passive double star	City (banking district)
3	Lippetal	March 92 June 92 Sept 92	≤4,500		Raynet Corp	Star	3 stages; specific designs
4	Leipzig	Nov 91	≤ 200	SELAlcatel System 12	Siemens	Passive double star	Residen-tial and business areas
5	Stuttgart	April 92	≤ 200	SELAlcatel System 12	SELAlcatel	Passive double star	Residen-tial and business areas
6	Nuremberg	Sept 91	≤200	Electro-mechanical	FAST (AEG, ANT, PKI)	Passive double star	Residen-tial areas
7	Bremen	Dec 91	742		Bosch Telecom	Star	Alter-native system to OPAL

2.6 COSTS

Telekom proved the suitability of FITL technology for providing narrowband and broadband telecommunications services within the framework of the BIGFON system trial at an early stage. However, for reasons of cost the system trial could not bring about an optical fibre access system to be directly implemented in a large-scale version.

Since then a worldwide drop in prices for optical and electrooptical components has taken place. However, the considerably reduced costs of the installation of a subscriber-tailored optical fibre access are still much higher than the costs of a comparable copper access — at least as regards the narrowband services. For this reason the planned establishment of a physical star-shaped FITL network in one step will continue to be unjustifiable.

The situation described above, i.e.

- the cost barrier which even in the long run cannot be handled by one network operator alone,
- the fact that there is no possibility of co-ordinating all network operators to create cost reducing demand,
- the lack of standardized and profitable services of broadband individual communication,

sets off the worldwide FTTH and FITL activities which aimed at circumventing the cost barriers. At the same time the long-term target is to continue to pursue a physical star-shaped FITL network.

Circumventing the cost barrier from today's point of view means:

- setting up low-fibre and thus cost-effective optical fibre structures;
- multiple and cost-effective use of system architectures;
- introduction of new transmission procedures (e.g. SDH) involving system changes;
- upgrading of established optical fibre structures and system architectures for enhancing narrowband services, for the timely provision of new broadband dialogue and distribution services and for the provision of not yet qualifiable and quantifiable services by external providers (open network provision — ONP);
- partial or complete transition to physical star-shaped structures with subscriber-tailored optical fibre access systems for all narrowband and broadband services.

This is a risky undertaking that must be carefully planned and carried out so that initial benefits do not have the opposite effect later or that the initial saving of costs does not result in cost avalanches later.

The cost analyses made and published by various network operators in the FITL area mainly relate to the more or less abstract calculation of 'initial costs' for establishing low-fibre structures (topologies) and the system architectures to be directly set on top. The results obtained this way are hardly comparable and hardly transferable.

For its cost analyses Telekom defined two model service areas that — with reference to the conventional copper cable provision — correspond to one cable distribution box area. Comparative plans were elaborated for these model areas comparing the conventional copper cable provision with the optical fibre provision on the basis of the companies' proposals submitted in the competition of concepts.

The model service area was a homogeneous residential area with mostly detached and semi-detached houses. In this area which was about 1.5 km away from the local exchange 182 POTS, 1 IDN access and 2 ISDN basic accesses were serviced.

Service area 2 was a block of buildings in the city centre with 12 multi-storey office and business buildings at a distance of 0.5 km from the local exchange. Here 150 POTS, 28 IDN accesses, 18 ISDN basic accesses, 2 ISDN primary rate accesses and two 2 Mbit/s fixed connections had to be provided.

As to cable installation and civil engineering costs all comparative plans used the same construction cost estimates per reference unit. The system costs of optical fibre transmission equipment were calculated on the basis of the companies' prices for the corresponding large-scale systems.

The cost comparisons led to the following results which are of great importance for Telekom.

- The costs of the optical fibre access systems depend to a great extent on the way the systems are connected to the local exchange. Connecting individual lines via interfaces costs much more than via multiplex interfaces. Because of this the demand for the provision of multiplex interfaces for bit rates $\geq$ 2 Mbit/s became mandatory. At Telekom's initiative, ETSI's SPS 3 Experts Group has meanwhile drawn up specifications for a multiplex interface (V5.1) that is suitable for connecting OPAL systems. This V5.1 interface will be implemented at the earliest possible date (1994).
- The economic efficiency of FITL systems as compared with conventional copper systems depends on the distance of the service area from the local exchange. The analysis of this (trivial) result for chosen concepts showed that the FITL systems, if connected via multiplex interfaces at a distance

of about 2 km, cost the same as conventional copper systems. Hence it will be necessary to enlarge the service areas for the purpose of optimizing the use of FITL.

- The costs of FITL systems differ structurally from the costs of equivalent copper systems. A comparison of the cost structures of physical star/logical bus PONs with that of the conventional copper system showed that the costs of the copper system in residential areas are almost exclusively, and in the business area to a large extent, made up of cable installation and civil engineering costs. By contrast, the costs of FITL systems are determined largely by the costs of the optical fibre transmission equipment. The costs of the optical fibre transmission equipment reflect the companies' insecurity in this area and this underlines the necessity for the network operators to establish viable demand.
- Integrating switched narrowband services and the broadband distribution service can contribute to considerable cost benefits.

Basically, a cost comparison should also include the cost of the operation, management and maintenance of equivalent copper and FITL systems. In this field, considerable cost savings could be achieved from today's point of view by using operation support systems (OSS), which would, for instance, avoid dispatching personnel if teleaction/telecontrol could be used. A quantification of this cost aspect is not possible at the moment.

2.7 CONCLUSIONS

Starting in 1993, Telekom will primarily use optical fibre technology at the subscriber line level of its telecommunications network in eastern Germany.

Telekom intends to connect 200 000 homes in 1993, 500 000 homes in 1994 and over 500 000 homes in 1995, via OPAL optical fibre systems.

Telekom is thus taking advantage of the unique opportunity that presents itself in eastern Germany and is the first operator worldwide to introduce optical fibre access systems on a regular basis to provide customers with narrowband dialogue services and radio and TV programmes.

The OPAL optical fibre projects described in this chapter are a cornerstone of Telekom's ambitious and pragmatic approach.

These projects make it possible for the manufacturers involved to produce the first optical fibre systems — adapted to Telekom's network — on a small scale, which will be purchased by Telekom in a public competition in 1993 and used cost-effectively in a few selected, limited areas in eastern Germany to provide the first 200 000 homes with services.

These projects will also provide important information to be used in drawing up specifications for standard OPAL systems to be purchased in an international competition. Starting in 1994, these systems will then be used in even larger areas at lower costs in eastern Germany and possibly also in western Germany.

Moreover, the geographical distribution of these pilot projects (Fig. 2.6) will enable the organizational units involved to obtain know-how early on and will thus support the implementation of systems on a large scale.

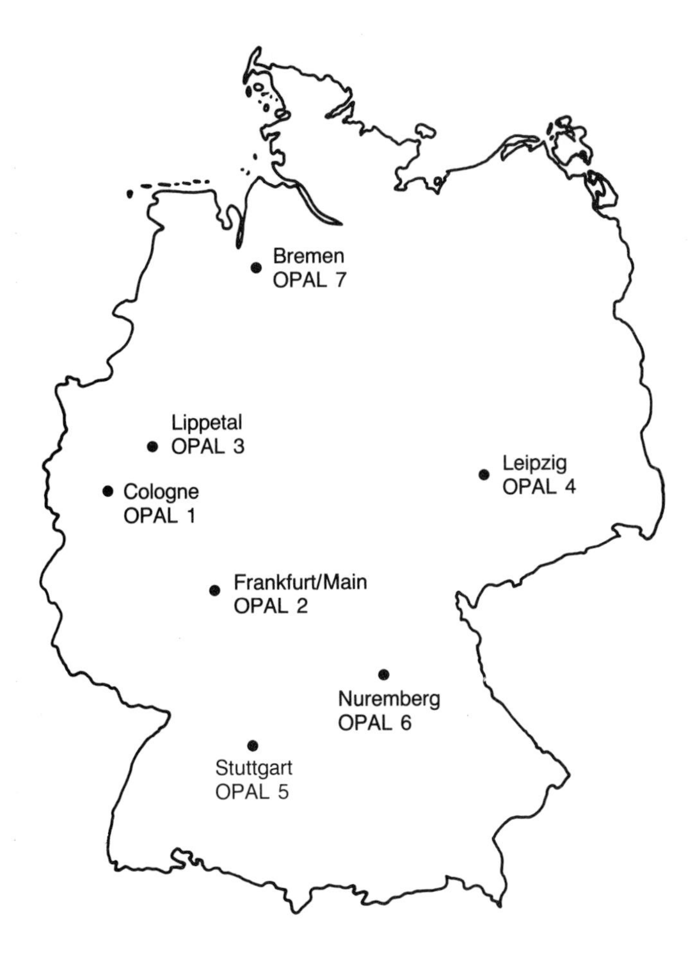

Fig. 2.6 Locations of OPAL pilot projects.

Telekom is making every effort to draw up specifications for standard OPAL systems to be used starting in 1994. For this purpose a special project unit was established at the Research and Technology Centre, formerly the Telecommunications Engineering Centre (FTZ), early in 1991. The project unit will prepare the basic conditions for an international invitation to tender for large-scale optical fibre access systems on the basis of international discussions open to all companies and experts. In 1992 all parties interested will be asked to comment on this specification. After its revision the technical terms of supply will be available for the invitation to tender at the beginning of 1993.

Telekom will, of course, participate intensively in the standardization of FITL systems at the international level. However, in Telekom's view the standardization of the optical fibre architectures should be limited in the medium term to what is indispensable and should only be extended step by step.

However, Telekom is aware that one network provider on its own cannot reach optimum results in this complex FITL area.

For this reason Telekom will arrange co-operation with other comparable European network operators and thus contribute to formulating an independent European position in the FITL area. This position will not only strengthen Telekom's platform of action but also contribute to reducing the costs of the desired FITL by concentrating demand in a European context.

REFERENCES

1. Tenzer G: 'Fibre to the home', R V Deckers Verlag, G Schenck, Heidelberg (October 1990).
2. Haist W: 'Optische telekommunikationssysteme', Band II: Anwendungen (December 1989).

3

RATIONAL AND ECONOMIC FIBRE SYSTEMS

J D Jensen

3.1 THE NETWORK OF THE FUTURE

The environment made possible by the vision of the broadband network of the future has been labelled by such terms as 'superhome' or 'electronic cottage'. The terms carry with them a sense of the magnitude of the potential change. In this vision, the typical residence of the past has been enhanced by the addition of new and amazing capabilities designed to improve the quality of life — video interchange is added to a routine telephone call, directed database access of video, audio, and textual information supplements the nightly fare of news and programming, purchases ranging from shoelaces to automobiles are now guided by true-to-life images of the products displayed on a high definition video monitor, and the home itself is monitored and directed in its use of energy and access.

All of these capabilities are envisaged as riding on an integrated digital network transferring hundreds of millions of bits of information between each user and a myriad of information sources. With most experts in agreement on the ultimate character of our communications future, the question becomes not where to go but how to get there. In the dearth of a guiding governmental directive, the renovation of the local loop plant of the USA is most likely to be determined by the market-place.

While estimates vary on the cost to implement a broadband network on a grand scale, the real question becomes: Who will pay for the improvement to the network? Under traditional rate-of-return regulation, public utility commissions scrutinized capital expenditures which were used to calculate the base on which the ratepayer bill was calculated — the guiding principle being to protect the public from excessive charges associated with lifeline telephone service. This approach is beginning to change slightly as rate of return is supplanted by incentives and price caps. While the public utility commission of Michigan, for example, now allows a portion of the earnings in excess of the prescribed rate of return to be used for network improvement, these amounts fall far short of the funding necessary to implement a ubiquitous network.

While the political climate within the USA remains volatile, recent events indicate that barriers established by the US Congress, the Divestiture Accord, and the Federal Communications Commission (FCC) are beginning to fall. These events, such as the removal of the restriction on providing information services and the FCC Notice of Rulemaking on Video Dial Tone, are pushing the former telephone monopolies towards competition and creating an environment which will foster the development of new applications and services.

The transition to an all-fibre broadband network will not be represented as a step function; rather, the path will likely be an evolution of network capabilities as services dictate. This approach is based on the assumptions that, first, the initial implementation of fibre to the subscriber will need to be accomplished at cost parity with the current methods and practices of the copper telephone network and, second, the implementation must provide a platform for future services. In other words, the deployment of fibre systems which are cost-equivalent to the traditional copper approach should not elicit concern from either the ratepayer or stock holder. In this deployment, however, the process replaces a medium of limited bandwidth, copper, with one of nearly unbounded capability, fibre. The implementation must be such as to preserve in the network the ability to tap and efficiently utilize that inherent broadband fibre bandwidth at some point in the future. Each of these two assumptions can be addressed more fully.

3.2 COST TODAY AND FLEXIBILITY TOMORROW

To understand the issue of cost parity with copper, one must first examine the complexion of the local loop. Rather than a single fixed cost target, the new fibre-based loop systems will need to meet a range of applications and approaches. In a loop survey compiled by Bellcore in 1983, the loop length varied from a minimum of 56 m to a maximum of 38 km (see Fig. 3.1). The mean was determined at 3.5 km and 75% of the loops were 4.6 km or shorter. Over the last two decades, central office consolidation extended the typical radius of a serving central office to 13 km. A carrier serving area (CSA) concept was developed to address the longer loops. The CSA implementation constructs remote electronic sites housed in controlled environment vaults, huts or cabinets which serve up to 2000 lines at a radius of 2.8-3.6 km. As is typical in urban and suburban areas, the majority of the growth occurs on the fringe areas. As this growth occurs, CSAs are identified and constructed using digital loop carrier systems, first introduced in the early 1980s, to address the expanding service needs (see Fig. 3.2).

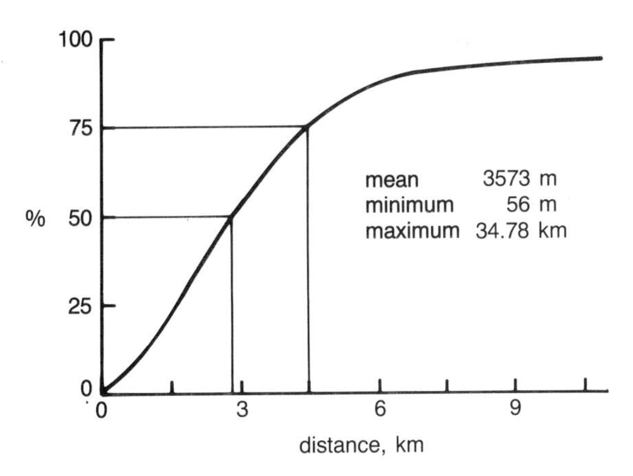

Fig. 3.1 Working length distribution residence loops (1983 Bellcore Loop Survey).

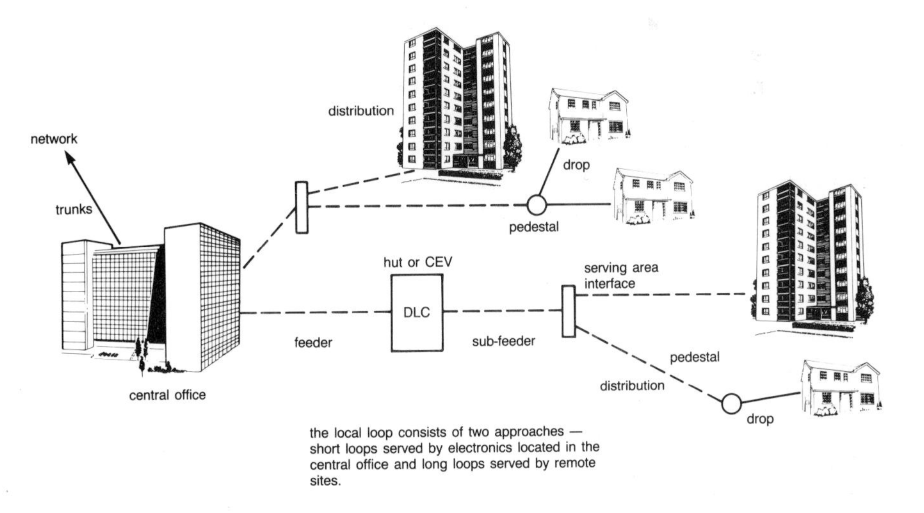

Fig. 3.2 The local loop consists of two approaches — short loops served by electronics located in the central office and long loops served by remote sites.

Current planning guidelines indicate that digital loop carrier is cost-effective for loops in excess of 3 km. Inside this distance (which again represents up to 75% of all access lines) service is provided directly from the central office and no remote electronics sites are deployed. Today, about 8% of existing access lines in the Ameritech region are served by digital loop carrier — a figure which encompasses both residential and business lines.

From a base, which in the case of Ameritech (comprising the states of Illinois, Indiana, Michigan, Ohio and Wisconsin) represents approximately 17 million lines, access line growth of about 2.5% occurs on an annual basis. This overall growth figure can be subdivided into two components — residential and small business growth which averages 1.2% on a base of 11 million lines while growth in the medium and large business sector is approximately 5% on a base of 6 million lines. While these two areas, residential and business, share some common services and capabilities, the focus on the market and technology aspects diverges. Business requirements are identifying capabilities such as diverse switching and routeing, LAN interconnect, and survivability. Emerging technologies such as SONET, 802.6 and switched multimegabit data service (SMDS) are being examined for these applications. In parallel with these efforts is the work proceeding on taking fibre to the residential and small business subscriber.

In addition to the new growth described, a continuous process is under way to refurbish and repair existing loop plant. This effort, typically called rehabilitation, is performed on an estimated 3% of the existing access lines on an annual basis, a number which is three times the residential new growth. Rehabilitation is driven by analysis of maintenance costs on a distribution area basis. Strict cost accounting of maintenance actions is maintained on each distribution area serving between 200 and 600 subscribers. Calculations of the anticipated maintenance savings versus the capital outlay to correct the problem are used to identify the leading candidates for repair. This effort also fine tunes the repair action identifying, for example, a specific splice area or terminal which constitutes the bulk of the problem. Thus a distribution area rehabilitation may only affect a portion (60% on average) of the loop plant in that area. A prioritized list is generated based on the return investment and funds are allocated.

Three areas of application have now been identified: growth served by digital loop carrier (DLC); central office-based (CO-based) growth served directly from the switch, and rehabilitation. Just as loop lengths span a wide range, the costs associated with the provision of services for those loops exhibit a large span with a simple rehabilitation job costing as low as $75 per line up to well in excess of $1000 for a new line a significant distance from the central office served by digital loop carrier. In between the cost of rehabilitation and DLC-based services is CO-based service. One recent job involved an up-scale condominium development approximately 2.4 km from the central office. The cost per line for this job was just over $400 per line. The key point is that there is no magic cost target for a fibre-based loop system in the US market; breakpoints do occur where the volumes begin to increase significantly. At a cost of $900 to $1400 per line, new growth in carrier serving areas could be targeted for fibre, as the cost decreases to between $400 and $800 per line, new growth directly out of the central office could be included. As the cost falls to the $200 to $500 range a significant portion of the rehabilitation market becomes available.

Historically, fibre-based distribution systems for deployment in the early to mid-1990s have been developed as extensions of or replacements for the digital loop carrier remote terminal. As such it is expected that, at best, these systems would only be cost-effective for the longer loops. It is ironic that digital loop carrier was developed to extend central office interfaces closer to the customer whereas, in the future, the fibre interface developed originally for digital loop carrier will be required to migrate to the central office. At this time, little interest has been generated among the major switch manufacturers or the appropriate standards bodies to address the short loop issue.

The inability to address all applications in a cost-effective manner may become a liability of fibre loop systems in the future. As new services are conceived for delivery over the fibre system, it will be important that large groups of potential customers be identified so as to facilitate marketing and distribution of the new service. If the only cost-effective application for fibre in the local loop is new growth on the CSA-based longer loops, this will result in a very thin smattering of potential customers over a wide geographic area which, in turn, will deter potential service providers.

It is also important to consider the transition from copper to fibre in the planning of future services. In order to attract new service providers to the network, it will be necessary to offer access to a large population — a feat made difficult if these potential customers must be served by fibre. One means to address the shortfall of potential fibre-based customers for higher bit rate services is to refine the transport capability of the existing copper network. Efforts are currently under way in national and international standards bodies to define a high-speed digital subscriber line (HDSL) which would provide up to 1.5 Mbit/s over two pairs of loop grade copper at distances up to 3.6 km. Thus with the appropriate introduction of electronics, services requiring a DS-1 could be provided over both the copper and fibre network. HDSL and a similar effort involving an asynchronous DSL operating at 1.5 Mbit/s in the dowstream direction only represent a synergistic effort to fibre in the distribution loop plant.

As has been indicated, cost is a critical factor in the realization of a sound fibre-deployment strategy. Without the existence of market drivers for services requiring fibre's bandwidth, copper will be difficult to dislodge from its position as mainstay for distribution. Early in the analysis of fibre in the local loop, the focus was placed on taking fibre all the way to the home. From a technological and operational point of view this is the preferred approach. Initial trials in the USA placed the optical network unit (ONU) on the side of the house and dedicated the function to a single subscriber. A single fibre or pair of fibres were dedicated between the customer's ONU and the serving electronics site. The most immediate problem with this initial thrust came from cost projections. A Bellcore study released in 1990 indicated that fibre to the home in a typical US housing development would cost in excess of $2500 in moderate volumes. Many other cost studies were conducted based on the premise that copper costs were inflating and fibre costs were decreasing with time; the result being that fibre did not prove in until the next decade. Clearly, volume was necessary to drive the costs down. At the same time, however, the costs needed to be driven down to generate the volumes.

One approach to break this impasse came from the analysis of the major cost components of the system. The ONU dedicated on a per customer basis represented the most significant cost factor in the system. Some of the items required in the ONU, such as the optical transceiver, would be subject to cost decreases as volume increased and technological improvements were made. However, a significant portion of the cost of the ONU could be traced to the simple provision of telephone service. These functions, which included the common control, battery, overvoltage protection, ringing, signalling, coder/decoder, hybrid and testing (simply known as the BORSCHT functions), represented a substantial cost when dedicated to a single customer in the provision of a single telephone line. Further analysis indicated that these costs increased only marginally when the system provided multiple lines. Moving the ONU to a point in the network, logically the easement in the front or back of the house, adds expense to the housing and construction but these costs are more than offset by the ability to divide the additional expense among multiple subscribers. The same Bellcore study cited above found the estimated cost for this shared approach to be less than $1400 per line. Thus, sharing of the ONU or fibre-to-the-pedestal (FTTP) became an arguable alternative.

The two options presented to telephone company planners involved waiting, perhaps ten years or more, for a combination of copper inflation and fibre cost decreases before fibre to the home (FTTH) becomes cost-effective, or implementing what appears to many a transitional FTTP product in two or three years. A consensus has developed among the regional Bell operating companies that FTTP is an appropriate step in the direction of a broadband network of the future. Indeed, FTTP may be necessary to generate the early volumes of the critical optical components required for the future cost profiles necessary to make FTTH a reality.

Bringing fibre within 150-300 m of the customer does not limit substantially the types of service offerings available in the future. Standard twisted wire pairs can transmit in excess of 10 Mbit/s over this distance, while placing shielded pairs in the drop could provide for transmission rates up to 100 Mbit/s. A coaxial drop could broadcast as many as 150 NTSC video signals or multiple DS-3 (45 Mbit/s) signals. The extension of a fibre drop with the replacement of the active electronics in the pedestal with a passive splitter or wave division multiplexer is also an alternative. Thus, as an interim step, FTTP does not seem to preclude any currently conceived future service.

Today, fibre proves cost-effective in feeder applications where as many as 2000 subscribers are served over the same feeder route. The intent of fibre in the local loop is to reduce the level of sharing, ultimately to the point where each customer is served by a block of bandwidth or, perhaps, a dedicated fibre. The level of sharing of the fibre system in FTTP represents a trade-off between cost and flexibility. As a larger number of customers are served by the ONU, the cost of the common functions on a per line basis can be reduced. At the same time, moving the fibre further from the customer and forcing the need potentially to share the bandwidth of that fibre among a greater number of subscribers reduces the flexibility in the implementation of future services. Therefore, the choice of the level of sharing must be a compromise between cost and future functionality. An early consensus was developed in the USA on a fibre system serving nominally four subscribers with up to twelve lines. Designed for placement at the intersection of four lots in a typical residential setting, the drop length between the ONU and the residence would be less than 150 m. The next step up in terms of sharing is the placement of an active pedestal serving four subscribers as well as two passive copper splice-only pedestals each serving four subscribers for a total of twelve. Cost analysis indicates that this additional level of sharing results in a 10% to 14% saving. In this scenario, planning would include plans to connect the active and passive pedestals via ducting to facilitate upgrade. Restricting the level of ONU sharing at four would cetainly be preferable; however, early deployment costs may dictate an alternative.

While much of the discussion has been focused on the typical residential subdivision, other applications must also be addressed. As a region, Ameritech purchases significant amounts of aerial copper cable as well as buried. This is an indication of the importance of finding an appropriate solution to the aerial distribution plant as well as buried. In general, serving sizes are larger for the aerial system and the system must be adapted for pole mounting. Special consideration must be given to cable access and hardware position. Size becomes a major factor when pole positioning is considered. Buried applications, however, must also be considered. In some subdivisions, all utility equipment must be out of sight. For telecommunications, this means that the ONU must be mounted in a hand-hole with access at ground level.

3.3 POWERING THE OPTICAL NETWORK UNIT

One of the most difficult problems faced in the deployment of fibre in the local loop is the powering of the end electronics. Early field trials of FTTH, in which the ONU was placed in a box on the outside of the home, found the equipment powered by a strategically placed a.c. outlet. Placement of the ONU, with the implementation of FTTP, at some point removed from the home complicates commercial power delivery and battery back-up.

Powering capabilities to support the function in a typical central office requires access to commercial a.c. power, the accompanying rectifier and converters to provide the 48 V required of telecommunications equipment, batteries and generators. Engineered to meet the expected load for normal operation of the office, the power system is designed to provide telephone service indefinitely to the customers served by that office. In the event of the commercial power failing, batteries with connections for back-up generators are in place to meet the power requirements. As equipment such as digital loop carrier was deployed outside the central office, guidelines were established with respect to the battery reserve with a minimum of eight hours required, assuming a per line usage of 9 CCS or 0.25 erlang. Coverage for a large portion of the expected commercial outages is expected. In the event that the outage continued longer than the battery reserve, portable generators can be utilized to extend telecommunications services.

Customers served by the newest fibre technology should receive the same or better level of service as their neighbours served by copper. In addition, as these systems electronics become more geographically dispersed, the need to address the power efficiency becomes an important life-cycle cost issue. While this aim is desirable for the entire network, it is especially obvious when considering the problem of FTTP. The overall objectives can be summarized as:

- provision of sufficient power to meet maximum system load requirements;
- demonstration of high efficiency in both power transport and equipment;

- provision of sufficient battery back-up to cover the 95 + % of a.c. power outages;
- a desire to provide for a means of indefinite power delivery.

Two components in a solution to meet the first three objectives are connection to a source of commercial power and the provision of energy storage, most typically batteries, to maintain system performance in the event of a power outage. First, access to utility power can come from two directions — the network or the subscriber/local. Second, placement of batteries must be considered.

In the delivery of power from the network, a common point of commercial power access is determined. This could be located at the same point as the serving electronics, such as a central office, hut or vault or located in or near the serving area. In either case, power is acquired for a large number of subscribers, and conditioned for transport to the served ONUs. Power transmission is most likely 48 or 130 V direct current allowing for runs of several thousand metres between the power utility source/converters and the ONUs. The capability of a network-powered system is limited by the voltage used as well as the gauge and length of the conductor (see Fig. 3.3). A separate overlay metallic network is necessary to support the fibre-based electronics. While it is not desirable to construct and maintain this additional network, network powering does provide a single-point control for power access, battery supply and maintenance, as well as an entry point for portable generators. This represents only a slight departure from current energy methods and practices for remote electronics.

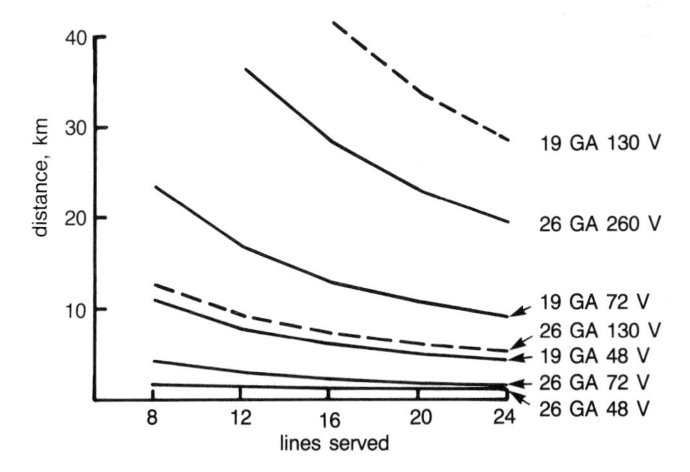

Fig. 3.3 Several combinations of voltages and loop lengths plotted on the number of lines served by an ONU versus the distance over which a network-powered system can operate.

An alternative source for commercial power can be derived from the power supplied by the utility to the customer. The power could come from a tap on the electric meter located on the side of the house from which direct current (d.c.) power is backfed on the drop to the shared ONU. Local powering has the advantage of providing the ONU with low-voltage d.c. over a short run. This approach allows for an all-dielectric plant reducing the susceptibility to lightning surges. In addition, loads up to tens of watts per subscriber can be powered, thus allowing graceful growth to all broadband services. It is not desirable to create the impression that one subscriber served by a specific ONU is paying more than his share in the powering of telecommunications services originating at the ONU; therefore all subscribers (or the utility connection at the subscriber premises) served would be involved in backfeeding power to their respective ONUs. Because of this, additional installation cost is also incurred because all premises served by an ONU would need to be equipped with the power access and backfeed function. It has the disadvantage of requiring batteries at or near the ONU such that extended commercial power outages would not result in loss of service since use of a portable generator is not feasible. Also, increased maintenance and remove/replace time is necessary owing to the high number of battery locations.

Since cost analysis indicates little difference between the two alternatives, the rationale for a decision may be based on objectives of greater scope than the technical questions. Some of the key issues which have been identified include:

- subscriber impact and perception of new technology;
- maintenance of a high level of service;
- technology to support long-term service evolution.

In all likelihood, both approaches will be used in FTTP deployment in the USA. For the Ameritech region, network powering will be the first choice. This allows for a transparent application of fibre technology especially in rehabilitation where established homes may not easily accommodate the power connection. However, in some new developments where aesthetics are of high concern, local powering may be applied to obviate the need for power housings located in the housing area. One issue which arises from the use of network powering is the evolution to future services. It is difficult to anticipate the power requirements for forward-looking services. In order to optimize the power system for delivery of today's expected services, insufficient margin may be left for future needs. This would require the network approach to include the appropriate 'hooks' to allow for conversion to local powering as future needs dictate. These hooks would allow for the local powering solution either to replace or to supplement the original network powering scheme.

Testing and diagnostics of fibre-based systems must also be considered in the design. Internal system diagnostics should include analysis of the performance of the system through digital loop-back testing as well as reflective and absorptive testing through the analogue hybrid of the customer's line card. Drop testing is also envisaged to assist in fault sectionalization. Tests indicating opens or shorts in the drop will trigger specific maintenance actions. Tests should also be included which sense the presence of high voltages and the presence of ringers on the loop. The technician should be assisted in the restoration of system performance in the event of a problem and should receive confirmation that their maintenance actions have resolved the problem.

3.4 THE CHOICE OF ARCHITECTURE NARROWS

While the level of sharing of the ONU has a primary impact on the system cost, the supporting fibre architecture has a secondary effect. The architecture's impact on cost is directly related to the amount of fibre required to provide service. As the fibre cable cross-section increases, costs associated with splicing and termination also increase. The two architectures receiving the most interest can be termed active star and passive star, perhaps better known as passive optical network. The active star, representing a point-to-point connection between the host digital terminal (HDT) and the ONU, requires more fibre and fibre support than the passive star. The trade-off of the passive star is the replacement of fibre for the passive splitters used in the distribution of the signal to/from many ONUs. Cost savings are quickly realized up to a point as the distance from the HDT to the ONU increases. One cost estimate indicated that savings of 8-15% were realized with the passive star when compared to an active approach. There still exists, as in the copper realm, a point where active multiplexing becomes cost-effective.

The passive star has other perceived advantages from a life-cycle cost and operations standpoint. Utilization of passive star architectures is expected to reduce the fibre complexity at the central office dramatically. For example, assuming an ONU serving four customers over unidirectional fibre facilities, a central office serving 20 000 residential lines would need to terminate 10 000 fibres for the active star while the passive star utilizing a 32-way split would require only 320 fibres. While an outage caused by 'unauthorized cable entry' may affect the same number of customers, the mean time-to-repair would be significantly less for the passive star owing to the reduced number of fibre repairs required. Covenants and restrictions in many cities make placement and access to remote structures difficult and expensive. In these cases, passive components can be used as an alternative to the placement of remote electronics sites.

A variety of cost and operational issues can be included in the decision of architecture; however, the choice should be determined primarily on the expectation of the future. As described earlier, the intent of fibre in the local loop is to create a platform for these future services. At the same time, it is important to defer as much of the costs related to these future revenues to the time when revenues are realized. Hence it is important to determine as far as possible the nature of the services to be provided and the time frame of their expected availability. The near-term expectation of ubiquitous high-bandwidth services such as LAN interconnects and video on demand contemplated in the introduction of broadband integrated services digital network (ISDN) would influence the network planner to select an active star arrangement. However, several factors weigh in favour of the passive approach. First, as described at the beginning of this chapter, the legal, regulatory, and competitive obstacles to new services are quite volatile. The management of the associated risk would favour the adoption of a conservative near-term approach which deferred market choices into the future when the environment is clarified. The same is true for technology choices. As new discoveries are introduced which advance the capabilities of the network, these can be applied to the existing network. Technology choices such as increasing the bit rate, increasing the optical budget through the use of optical amplifiers, adding wavelengths or fibre all represent viable alternatives which can 'compete' for the application of network upgrade. Another important consideration is the time value of money. It is preferable to match the infusion of capital into the network with a matching revenue return. As a result, considering the lack of a business case which supports the near-term ubiquitous deployment of BISDN, the passive star represents the archiecture choice for the local residential and small business loop.

One area which can be more fully explored is the technology risk associated with the deployment of a passive star system. The services requiring transport on the loop plant of the future could take two distinctly different directions. One option would be a service mix dominated by a combination of telephony and broadcast analogue or digital video services. In this case, the point-to-multipoint nature of the passive star is ideal for the distribution of broadcast information. Another view would indicate, given the current cost effective delivery of broadcast video over a tree and branch coaxial plant, that the passive star is preferred since the delivery of broadcast information over many point-to-point connections does not exude economic or technical wisdom. Alternatively, the prevailing service mix of the future could be oriented towards switched digital services with a high degree of interactivity. In a passive star system a variety of alternatives can be considered for service delivery. The split ratio developed for narrowband services can be reduced by introducing feeder fibre which taps into a later stage of the splitter array. The system could then continue to provide telephony to the original number of subscribers with broadband services dedicated to smaller groups based on the number of splits remaining in the path of the broadband service. As with telephony, the headend for new services can be shared over a number of subscribers dictated by the economics of the service.

Consideration should also be given to the capability of the ONU. It is desirable to preserve the investment in this equipment as long as possible. The ONU backplane should be designed to accommodate an increased range of bit rates. One option would allow the upgrade to DS-1 1.5 Mbit/s service to be extended from the ONU with only module replacement. The ONU should also contain the hardware necessary to allow inclusion of wave division multiplex devices in the fibre path for future upgrade. Accommodation for a variety of drop types such as fibre, coaxial cable and shielded twisted pairs should also be considered. The ONU will be the most visible part of the network to the customer. As such, it is important that the unit be as unobtrusive as possible. At the same time, it is going to be placed in the portion of the network which is subject to a considerable amount of abuse ranging from automobile encounters to permanent second base for backlot baseball.

3.5 TECHNOLOGY'S MARCH WILL CHALLENGE SERVICE PROVIDERS

The message should have become apparent that the intent of fibre in the local loop is to create a flexible platform for both present and future services. The local loop represents 90% of the investment made by telephone operating companies. The fibre architecture together with the optical network unit represent the basis for communications into the next century. The introduction of fibre as the medium of choice in the local loop will radically change the nature of services and capabilities in the future. Perhaps the most notable change will be the transition from the administration of copper pairs to the manipulation of bandwidth. The supporting methods and procedures for the network will, by necessity, need to be adapted to this new approach. In the long term, this change will also profoundly affect the fundamental business of the local telephone companies. In the past, the most significant investment has been in the passive cables connecting subscribers with the network. The investment direction will now change with the functionality on the network determined by the electronics placed at the periphery of this new fibre network. This change is coming at a time when increasing competition is appearing and substantive changes are occurring in the structure and complexion of the organization. Fibre in the local loop is among several other key initiatives which will introduce changes to the fundamental business structure of the telephone company.

4

FIBRE IN THE ACCESS NETWORK

K A Oakley, R Guyon and J Stern

4.1 INTRODUCTION

As optical fibre is now used for virtually all growth in the core transmission network, expectation has been growing over the last few years that the 'last mile' of the local access network would be the next step.

The access network however is proving a tough challenge. Fibre has traditionally been economic in the core network because of its high capacity and long-range capability. The fact that all new circuit capacity is digital greatly assists core fibre economics. In the access network, distances are short, capacity needs are currently low (most customers have only one line) and, although the exchange end is increasingly digital, the customer end is an analogue telephone interface dating back to the last century. However, copper is increasingly expensive, has quality problems, has long lead times for service provision and cannot support future broadband services. All of these are problems that single-mode fibre can help solve.

There have been many false dawns for access fibre. There have already been major fibre successes such as London's City and Docklands fibre networks (CFN/DFN), new megastream provision is increasingly over fibre, and the fibre in the access network (FAN) programme to major customers has already started. The mass use of fibre however awaits new technology

---

[1] This chapter is reprinted, with only minor amendments, from an article which first appeared in British Telecommunications Engineering Journal Vol 10 Part 1 April 1991 and is reproduced by kind permission.

that will reduce equipment costs and increase the applicability of fibre to all customers not just a few major ones. Such technology is now in view with BT's Martlesham Heath Laboratories very much in the forefront of a major international drive to get economic solutions available for fibre into the loop (FITL), a term widely used in North America.

Fibre will make a major impact on the access network in the 1990s, presenting a significant challenge to all those working in the access sector.

4.2 BASIC SYSTEM DESIGN

A major difference between copper and fibre is that, while a telephone can be directly connected to a copper pair and power fed over it at the traditional 50 V, in a fibre system the pulses of light sent down the fibre need conversion to a copper pair to serve a conventional telephone. This requires a box of electronics at the customer's end and at the exchange. Power must be supplied locally at each end. However, while the capacity of a single copper pair is normally a single exchange line, or with future electronics up to perhaps ten lines, that of a single fibre is initially hundreds of lines. With an upgrade, thousands of lines and TV channels can also be provided.

Figure 4.1 illustrates a typical early system, using equipment derived from junction network use. Copper pairs leaving an exchange are taken via the main distribution frame (MDF) to a primary MUX (PMUX) in the exchange. This converts the analogue signals on 30 pairs to 30×64 kbit/s digital signals and multiplexes them up to a 2 Mbit/s signal. Four 2 Mbit/s signals are multiplexed together up to 8 Mbit/s in a higher-order MUX (HOMUX) and then passed to a line terminal unit (LTU). The LTU converts the 8 Mbit/s electrical signal to a series of light pulses on the fibre using a laser or, for shorter distances, a light-emitting diode (LED). Light is transmitted on one fibre and received from the distant end on another.

At the customer end, the network services module (NSM) containing an identical LTU and HOMUX feeds 2 Mbit/s electrical signals to a PMUX. The PMUX has line cards which convert the 64 kbit/s digital signal to the familiar 50 V analogue telephony interface which can then be fed over a copper pair to the telephone. The NSM includes a local mains power supply with battery backup. By changing line cards, a variety of services such as analogue private circuits can be supported. The 2 Mbit/s bearers can be used directly to support megastream and ISDN 30 services.

When the exchange is digital (System X or AXE10), considerable cost savings can be made. A digital exchange handles calls in the form of 64 kbit/s signals, converting them at the periphery of the exchange into analogue format

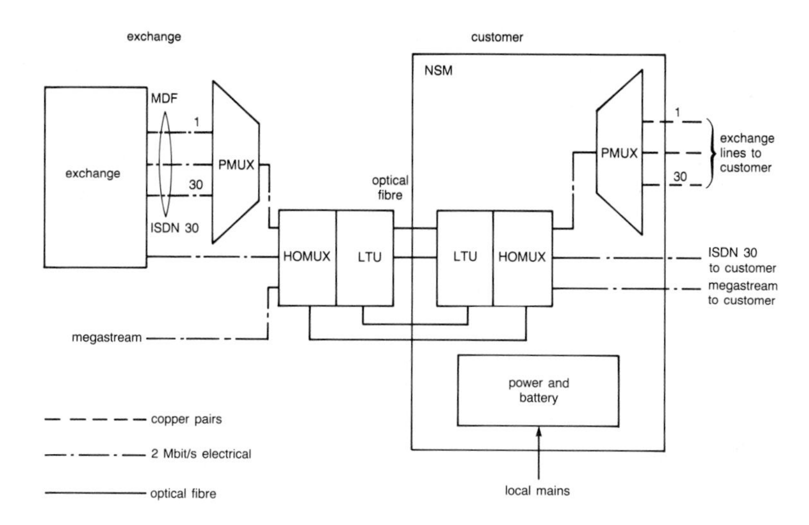

Fig. 4.1 Typical early system using back-to-back multiplexers.

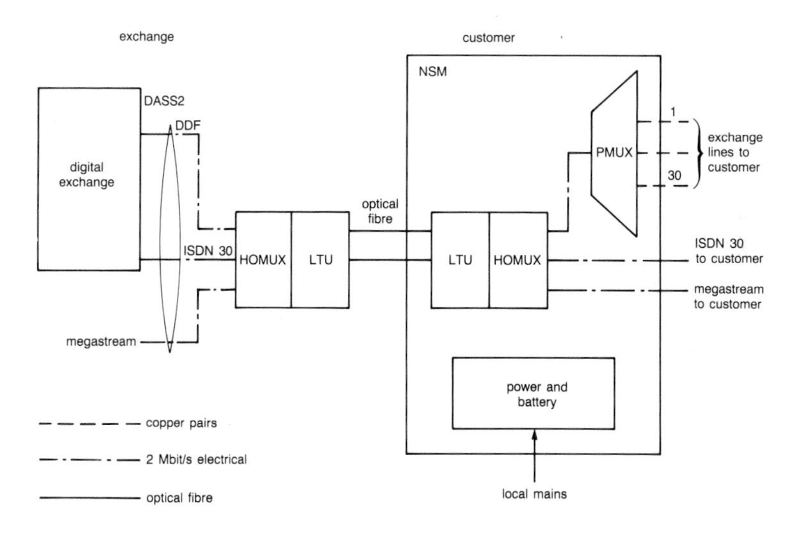

Fig. 4.2 Integrated PSTN using DASS2 interface.

using a line card. In the simple case of Fig. 4.1 therefore, a line passes through three line cards between exchange and customer. As Fig. 4.2 shows, only one line card is required if the exchange can output 30 lines in the form of a 2 Mbit/s electrical signal. A 2 Mbit/s interface using digital access signalling system No 2 (DASS2) common-channel signalling is now becoming available on BT's digital exchanges. It can support telephony, thus allowing BT freedom to purchase transmission equipment from a variety of manufacturers independent of the exchange type. An open interface such as this is an ambition of many Telcos in Europe and the USA, but BT has one of the first to enter service.

4.3 FLEXIBLE ACCESS SYSTEM

The flexible access system [1] (FAS) (Fig. 4.3) represented an evolution of the basic system in Fig. 4.2 above. To increase reliability the fibre system was duplicated using alternatively routed fibres where possible. The concept of a service access switch (SAS) was added to enable private circuits to be connected from one customer to another without expensive conversion back to analogue and connection via an MDF. A number of 2 Mbit/s bearers, each carrying up to 30 private circuits from a single customer, are fed into the SAS which then provides digital cross-connection between circuits at 64 kbit/s. Thus, instead of by the traditional MDF jumper wire, private circuits

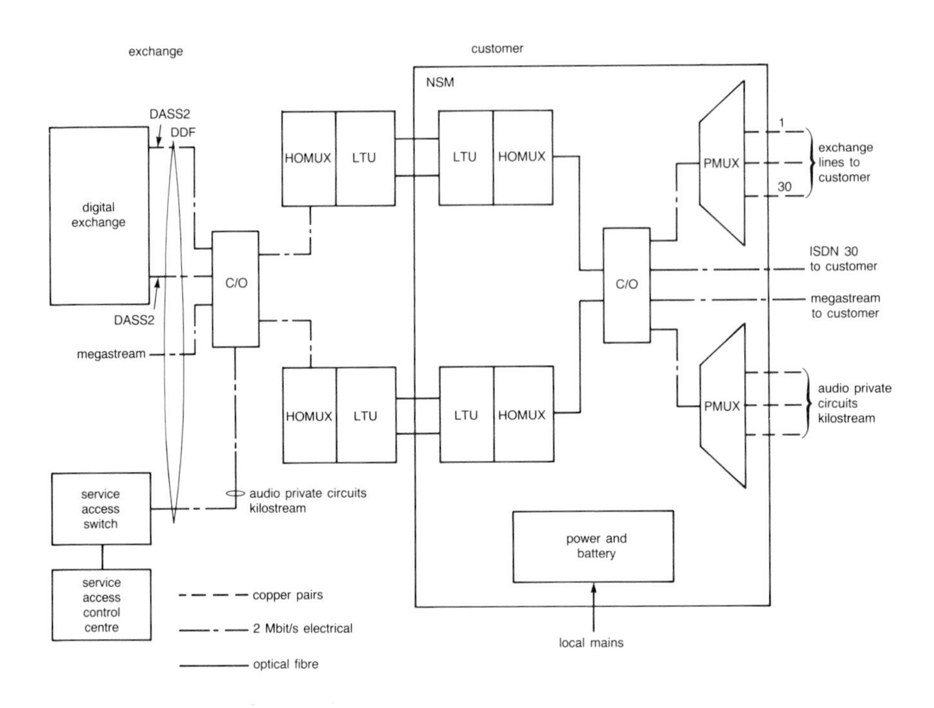

Fig. 4.3 Flexible access system.

can be routed electronically via a computer system under the control of staff in the service access control centre (SACC).

During the late 1980s, BT planned an extensive programme to use the FAS system to serve major customers (typically over 50 exchange lines). Although the design owes its origins to trunk and junction equipment, considerable development was needed to reduce the cost to a point where it was economically attractive. To recover that research and development, manufacturers required commitment to large-volume purchases. BT began to have worries about the size of the capital investment programme needed and decided that it was not the right time to make a commitment to that approach when technology internationally was changing rapidly. The FAS concept does however live on in the form of the CFN/DFN in the FAN programme.

4.4 CITY AND DOCKLANDS FIBRE NETWORKS

Stimulated by the advent of the Stock Exchange's 'Big Bang', the City fibre network and docklands fibre network were created to meet the increasing demand for the provision and rearrangement of private circuits within the city and docklands areas of London. The primary requirements of both system designs were:

- that they support the full range of electrical interfaces traditionally supported by the DealerInterlink service over the copper access network;
- that they were capable of accommodating the high level of churn endemic to the financial sector.

The traditional FAS architecture (Fig. 4.3) described earlier was selected. Point-to-point optical line systems provide secure, duplicated access to the serving sites where access to the core network or one of the network's eight SASs is provided.

In all there are four serving sites with a potential capacity of over a quarter of a million private circuit ends and 60 000 public switched telephone network (PSTN) accesses.

The network has been introduced in phases with each award of contract having been made as a result of separate competitive tendering exercises. The initial two phases were entitled city fibre network phase 1 and 2 (CFN1 and CFN2), while the third and fourth phases were given the names docklands fibre network (DFN) and Canary Wharf. Currently, analgoue PSTN is only supported by DFN and Canary Wharf.

Today the network is providing more than 20 000 channel-ends to several hundred customer sites featuring services such as audio private circuits, megastream, ISDN 30, kilostream and international 2 Mbit/s links.

As for the future, planning has now started on a scheme to enable the deployment of systems across other areas of London. This has been made possible following the start of the installation of a fibre infrastructure in the access network in London, initially focused on meeting future demand for 2 Mbit/s services. It will support both current and future fibre access technologies. The fibre access equipment will be deployed from customer sites to the serving exchange with the potential to link into existing CFN and DFN networks. The equipment will be deployed against criteria designed to maximize the benefits for both the access network and customers; thus these systems will be targeted on high 2 Mbit/s (ISDN 30/megastream) and private circuit users. The size and modularity of the current equipment means that it will be deployed in the larger sites, those with several 2 Mbit/s services, in particular greenfield sites. As new technology becomes available the range of products and services will be extended.

4.5 MEGASTREAM

The megastream service is currently provided using standard junction transmission equipment. Network terminating units are sited at each end of the serving section to provide the alarm and test functions required by the megastream maintenance duty at the XStream service centre (XSC). Additional equipment is provided to transport the alarms over either an auxiliary data channel or private circuit. The ISDN 30 service is delivered using the megastream equipment but it is specially configured to relay the alarm reports to the repair service centre (RSC) instead of the XSC.

Currently, 50% of all megastream growth is provided over fibre systems. The 4×2 Mbit/s system is, by far, the most popular delivery mechanism for fibre megastream. The number of sites demanding large numbers of megastream circuits is steadily increasing and this is resulting in more frequent deployment of 16×2 Mbit/s systems. The use of transverse-screen copper cable is now restricted to single megastream circuit customers, and fibre is used for most new sites. Thus fibre is accounting for a steadily increasing slice of the whole market.

Of late, the benefits of creating a generic 2 Mbit/s delivery mechanism with a single, fully specified alarm interface have become apparent. Work is currently under way to define the system architecture and the level of functionality required in the line system and NTU. The generic system will use the transmission network surveillance system (TNS) for the collection and processing of alarms and performance information for the 2 Mbit/s bearer and the service specific information. TNS will also provide control of the loopback facility resident in the NTUs.

4.6 INDIVIDUAL FIBRE SCHEMES

Individual fibre schemes (IFSs) comprise a variety of different transmission systems and primary multiplexers. By far the most common variant is the back-to-back (Fig. 4.1) connection of PMUX, employing integrated optical transmission equipment, delivering PSTN and analogue private circuits. Most of these installations site the customer-end PMUXs in environmentally friendly communications rooms but a 'street-hardened' version has been tested in Manchester with great success.

IFSs are expedient systems designed to meet localized unforeseen demand. Only when these schemes are considered economic or necessary for technical reasons are they approved. Because of their low penetration and 'special' status, IFSs survive operationally without the considerable backup organization and procedural changes necessary to accommodate large-scale roll-out programmes such as FAN.

4.7 FIBRE IN THE ACCESS NETWORK

Fibre in the access network (FAN) is a programme of phased fibre equipment deployment into the access network. The current architecture for FAN is, basically, as shown in Fig. 4.2, but with the additional capability of supporting a range of services through a variety of remotely sited primary multiplexers.

Primary multiplexers using DASS2 signalling are linked to the digital local exchange (DLE) by optical transmission systems to provide PSTN. Access to the DLE is at the primary rate and this removes the need for the analogue line cards at the exchange. Capacity at 2 Mbit/s is also provided over FAN enabling the delivery of megastream and ISDN 30. By employing the appropriate primary multiplexers in the network services module (NSM) and/or at the customer's site, analogue private circuits and kilostream plus can also be supported.

The procurement of the first tranche of FAN equipment has been based on a 'bottom up' approach to the economic modelling, potential customer sites having been identified through a strategic marketing approach.

Approval for a trial of the 'first tranche' FAN equipment has been obtained. The trial, near Bristol and in London, began in July 1991. Roll-out of FAN equipment started late in 1991.

4.8 TELEPHONY OVER A PASSIVE OPTICAL NETWORK

A key problem with the early systems described above is that they are only economically suitable for large customers. Since smaller customers are still served via copper, both fibre and new copper cables have to be provided in parallel to a growth area with therefore little saving in cable or duct costs. If all sizes of customer, and all services can be provided over fibre, there is a significant snowball effect on cost savings in cables and ducts.

Another problem is that of high LTU and HOMUX costs, which, although acceptable when shared over the 100 lines of a major customer, become very expensive when the same fixed equipment cost is divided over only say ten lines. Similarly, the use of four dedicated fibres from exchange to customer becomes expensive when that cost is shared over only ten lines.

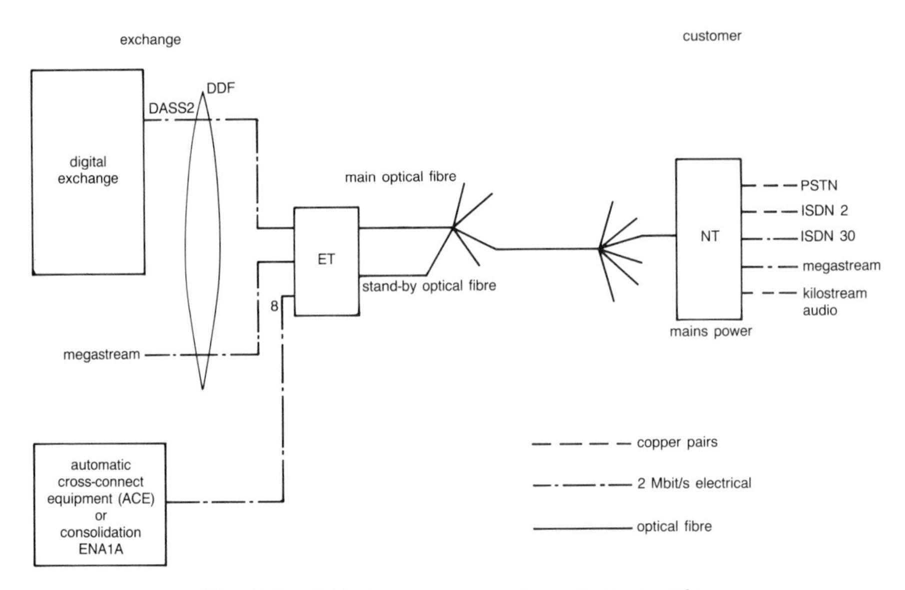

Fig. 4.4 Telephony over a passive optical network.

BT researchers at Martlesham Heath have produced what is now widely anticipated to be a solution to this problem in the next few years. BT's telephony over a passive optical network (TPON) system (Fig. 4.4) features a single fibre from the exchange that is split at the cabinet into say four fibres and then each fibre is split again into say eight more fibres. The resulting 32 fibres can serve 32 individual customers. Such a fibre network is now referred to internationally as a passive optical network (PON). Network terminals (NTs) on the end of each fibre feature new-generation low-cost opto-electronics (lasers and receivers) and synchronous higher-order multiplexing to dramatically reduce the cost and size of the LTU/HOMUX function. The use of splitters to enable a single exchange-side fibre to serve several customers reduces both fibre and exchange LTU costs.

The NT could take several forms.

- Business TPON — for business applications, it would be a small briefcase-sized wall-mounted box providing 4 to 30 lines on the customer premises with power supplied by the customer. It could be fitted in the basement of a large building as with FAN or small units could be dispersed around the customer's premises to deliver services directly. In the latter case, a pencil-sized splitter in the customer's basement would collect and distribute signals from each of the units to a single D-side fibre.
- Street TPON — for residential customers, a similar multiline unit fitted in an underground joint box at the distribution point (DP) would provide service to a number of customers over conventional copper drops. A range of power supply methods are being researched including power-feeding several DP-located units from a power supply cabinet or backfeeding power from one or more customers.
- House TPON — in this form, the NT is a small box about the size of a double power socket mounted on the customer's premises. It is locally powered from the customer's mains. A splitter is used at the DP to serve say eight houses from a single D-side fibre. Today's high opto-electronics costs mean that street TPON is likely to be used initially because those costs are shared over a number of lines. House TPON will probably not be economic unless there is a market for broadband to help carry the extra cost.

4.8.1 Time-slot routeing features

The experimental version of TPON developed by BT Laboratories accepts up to eight 2 Mbit/s streams from a digital exchange (using DASS2 signalling) or from special services such as the kilostream network. They are converted into 240×64 kbit/s time-slots broadcast over the fibre PON to up to 32 NTs. Each NT is instructed which time-slots to select by a network management centre, via a discrete housekeeping data link over the fibre. The NT transmits back to the exchange terminal (ET) using time-division multiple access (TDMA). This technique features the sending of bursts of data from the NT which are timed so that the converging bursts passively interleave at the cabinet and DP splitters. Another inbuilt feature of the TDMA bit transport system (BTS) is that each of the time slots can be freely allocated to any

NT. Thus traffic for a particular service, say kilostream, can be groomed from a number of NTs and consolidated into a single 2 Mbit/s stream from the ET into the automatic cross-connection equipment (ACE) network. This offers considerable benefits over earlier systems which incur high costs by taking traffic from only one customer and often under-utilize exchange or ACE ports.

The term TPON, although initially applied to the experimental system developed by BT Laboratories, is now beginning to be applied more generally to describe all PON systems that feature the use of TDMA techniques to control bit transport across the access network.

4.8.2 Maintenance features

The TDMA approach involves measuring the time of flight of the light pulses and thus the ET-NT distance, accurate to about 5 ns in 100 ms (equivalent to 1 m in 20 km), as well as measuring the optical loss to ± 1 dB. As this process is automatically carried out on installation of each NT and several times per second subsequently, it provides a very useful acceptance test and subsequent early warning of service failure as well as a means of locating some faults. The housekeeping data link from each NT can also be used to pass various alarms and test loop-back commands and to control a simple lost head which can test the copper drop.

In order to improve reliability, the ET and exchange-side fibre to the cabinet splitter are duplicated (Fig. 4.4). This is achieved by connecting a stand-by exchange-side fibre (routed over an alternative cable if feasible) from a spare splitter outlet back to a stand-by ET. In the event of ET failure, the 2 Mbit/s interfaces are switched automatically to the stand-by ET. Since the ET is shared by 240 lines, the cost implication is minimal. Each NT serves a maximum of 30 lines and is thus not normally duplicated. Larger customers are served by multiple NTs.

A key concern in the past has been to keep track of equipment, for example, where it is installed, the capacity available for particular services, and occasionally to track down a particular mark of card to facilitate its replacement because of a known defect. It is intended that production TPON systems will include a remote inventory system that will enable staff at the network management centre to discover instantly the type of cards fitted in a particular customer's NT and their serial numbers.

4.9 BROADBAND OVER A PASSIVE OPTICAL NETWORK (BPON)

The TPON system operates at a wavelength (or colour of light) of 1310 nm. By adding other wavelengths, it is possible to add a wide range of new services in future simply by adding terminal equipment to specific customers without the need to lay more fibre cables [2] (Fig. 4.5). These broadband services could be entertainment services such as cable television, although under current Government proposals BT will not be able to enter this market for up to ten years, or they could be business services such as high bit rate data, video-conferencing or video telephony.

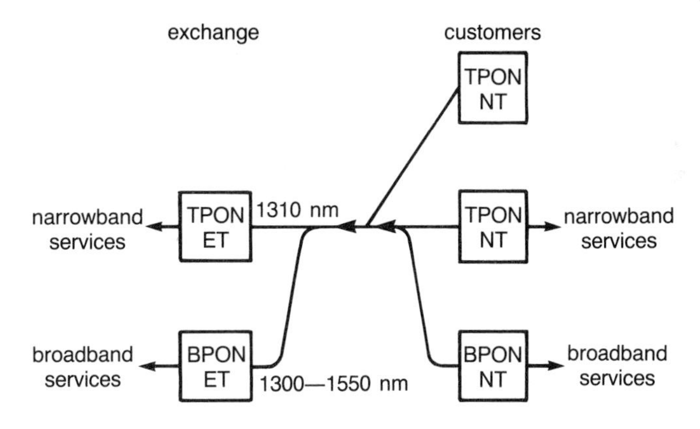

Fig. 4.5 Broadband over a passive optical network.

Several different modulation techniques have been demonstrated over a PON fibre architecture. One approach is to use a different electrical frequency (all on a single optical wavelength) to and from each subscriber. This frequency-division multiplexing (FDM) approach has the benefit that many different types of signal and bit rates can be used. One early system offers 16×2 Mbit/s paths, each 2 Mbit/s modulating a different frequency. Bit rates not meeting the CCITT standard, such as typical local area network (LAN) rates of 10 Mbit/s, can also be transported; this gives a convenient and cost-effective network alternative to specially laid cables from customer to customer.

Another modulation scheme uses TDMA at 150 Mbit/s, rather than 20 Mbit/s with TPON, to offer high bit rate asynchronous transfer mode (ATM) services to data customers. The APON system (ATM over a PON, see Chapter 12) would allow customers who want to send a burst of data of anywhere between 0 and 150 Mbit/s to access instantly that amount of capacity from the network just for the period they require it, rather like packet switching.

More futuristic upgrades include the use of high-density wavelength-division multiplexing (WDM). Techniques are available now to use say 4-5 wavelengths in the 1300 nm and 1500 nm optical windows with a spacing of 20 nm between wavelengths.

Research has shown however that over 30 wavelengths could be added at 1 nm spacing using tuned laser transmitters and electronically tunable optical receivers (see Chapter 15). Thus each customer on a 32-way split network could send and receive their own wavelength. Each wavelength could support up to say 2.4 Gbit/s although in practice a realistic application in the late-1990s in more likely to be a 150-600 Mbit/s broadband ISDN (BISDN) service.

Optical-loss consideration restricts the number of ways a fibre can be split; each two-way split introduces 3.5 dB of loss. However, erbium fibre optical amplifiers now offer the prospect of a simple-to-use device which greatly increases potential split rates. One 'hero' experiment has already demonstrated the transmission of 384 high-quality digital TV channels by using 2.4 Gbit/s on each of ten wavelengths from a single source to 39 million customers using just two amplifiers. The key significance of amplifier technology is that in the early days of a speculative new broadband service it allows the service to be offered to customers on many existing PONs for a low initial cost. Most costs are in the customer's NT and thus only incurred on demand.

Advanced optical technology thus offers the prospect of a wide variety of broadband services that can be economically added to an already installed PON network.

4.10 INTERNATIONAL PERSPECTIVE

Internationally, BT is in the forefront of the fibre-in-the-loop revolution. It is believed that the 15 000 fibres terminated on Baynard House in the City of London are a world record for any single exchange. A number of countries use FAS-type technology in a small way, usually to very large customers. Some use fibres to street multiplexers with copper drops of 500 m to 1 km to the customer. The USA in particular has around 10 million lines fed over digital loop carrier systems, many served over fibre links between the exchange and the remote multiplexer site. However, because of the greater overall length of the local loop in the USA, the remote multiplexer site (often up to 4000 lines and underground) still has a copper distribution network beyond it of up to 4 km in length. The USA remote multiplexer site is therefore often where the local exchange is situated in the UK (and Europe).

World interest in using fibre in the access network is growing rapidly. BT has been actively promoting the passive optical network concept besides other telephone companies and leading international manufacturers in the interests of gaining rapid consensus on a common approach. Although systems may vary locally, commonality in certain key components such as splitters, lasers and optical receivers will lead to low costs for all Telcos. This approach has been very successful, with Telcos in the USA, Europe, Japan, Australia and New Zealand now conducting research and development, producing specifications and setting up trials. The Deutsche Bundespost anticipates that the new access network being built as part of the overall national infrastructure programme in eastern Germany will be fibre- rather than copper-based from the outset.

4.11 BISHOP'S STORTFORD TRIAL

A trial of business, street and house TPON is being conducted at Bishop's Stortford in East Anglia (Chapter 9) [3]. The trial is intended to demonstrate the technology and investigate a number of practical issues. It uses fibre to support both telephony and 16 to 18 channels of cable television, the latter using a simple FDM approach based on the use of standard off-the-shelf satellite-TV set-top boxes [4]. A special two-year trial licence has been granted by the Government to allow BT to provide CATV over its main network.

The trial also features an alternative technology to TPON — broadband integrated distributed star (BIDS). In this approach, active electronics are used in the street in lieu of splitters to multiplex signals to and from several customers on to a single exchange-side fibre. This approach is not favoured in the long term because of its higher cost and the physical size of the cabinets used for the electronics.

4.12 FORECASTING THE FUTURE

A major concern facing BT is that it is continuing to invest record sums in a copper access network to meet today's narrowband demand while the future may lay in providing broadband services which need fibre.

All telecommunications investments are driven by forecasts of future demand particularly in the access network where major new copper cables can take up to 18 months to plan, provide duct, and lay and joint cables. Forecasts have historically never been accurate, particularly at the cabinet or DP level where unforeseen local events often weigh more strongly than national trends. In the future, a number of factors will combine to make forecasts even more unreliable, notably:

- as the network reaches saturation (that is, virtually all households have a telephone), residential growth becomes mostly demand for second and third lines, predicting just who will ask for such lines being very difficult;
- the effects of network competition;
- migration of businesses to ISDN 2, which provides effectively two lines over a single copper pair;
- migration of larger customers to ISDN 30 (2 Mbit/s), which requires fibre rather than its equivalent 30 exchange lines that would have been fed over 30 copper pairs.

BT's objective is to meet service on demand. In a copper environment with 18-month lead times, spare capacity must be available at the point of sale (the DP). The alternative is expedient pair provision or diversion on demand which generally is extremely costly. BT is therefore caught between expensive speculative initial investment in pair capacity or expensive expedient provision.

4.13 FUTURE VISION

The solution may lie in the use of fibre and pair-gain over copper. Unlike copper, where most of the cost is in a fixed initial cable investment, fibre schemes have only around 20% of the eventual network cost in the fixed fibre cable and duct — 80% is in the terminal equipment. The aim must be to provide only just the right amount of terminal equipment to meet customers' immediate needs and to be able to increase or change it within a few hours or days on request from the customer.

Most service provision would therefore become a simple task of remote commissioning and testing a pre-provided line card from a customer service system (CSS) VDU many miles away or of dispatching staff to the customer to fit another line card or extension box on the wall. With careful design and organization both could become simple, rapid tasks. Diverting cables, chasing spare pairs, with all its knock-on fault liabilities, would become a thing of the past.

The future network vision therefore is a network where exchange side growth by large copper cables has ceased and is replaced by a steadily growing thin veneer of fibre to cabinet locations and hence on to business customers and some greenfield residential situations. Exchange-side fundamental schemes that today speculatively place multiples of 100 pairs to exhausted cabinets (or cabinets passed by which exhaust in the next few years) might in future place two fibres in lieu of each 100 pairs at the cabinet location. In cabinet (and E/O) areas serving business growth, these fibres would be extended to business customers and their new growth taken up via fibre. In cabinet areas with primarily residential growth, the first option would be to extend fibre to existing businesses, converting them to fibre and using the pairs made spare to satisfy residential growth. Alternatively, street TPON or electronic pair-gain equipment [5] fed over existing copper pairs can be used.

Such a concept would aim:

- to maximize the use of existing copper spare pair capacity;
- to limit speculative fixed investment;
- to focus growth progressively, particularly business, on to fibre systems;
- to build up a fibre infrastructure for new broadband services.

Fibre will not replace copper in the next few years; the issue is to get the right balance between the two. The best of both technologies can be obtained by fully utilizing the existing copper investment for the bulk of the embedded network, and deploying fibre for growth and areas where most change is likely, such as businesses.

4.13.1 Fibre ready

Full implementation of the vision above is three or four years away, but there are steps that can be taken now to ready the network for fibre. One approach is to build up the provision of fibre spine cables (up to 96 fibres) in the exchange side. These can be justified today for the provision of 2 Mbit/s, but a key aspect is to ensure that they are sensibly dimensioned for future needs rather than just the immediate demands of perhaps a single customer. Another approach is to ensure that copper distribution-side cables being provided now allow for future installation of fibre. One method being researched is the inclusion of a small plastic tube (say 5 mm diameter) in the centre of all new distribution-side cables of perhaps 20 to 100 pairs. This would facilitate the easy subsequent blowing-in of a bundle of up to eight single-mode fibres [6]. The use of tubing rather than actually providing fibres from the outset reduces cost, minimizes the need to both immediately retrain and expensively re-equip today's copper jointing staff, and reduces extra on-site installation time to a minimum. It allows most costs and decisions on fibre layout, position of splices, splitters, etc, to be deferred until needed.

4.13.2 Challenges

Many challenges must be faced before the vision can be delivered.

- Equipment must be developed and manufactured that is low cost yet meets the exacting performance targets required in the access network.
- Field operations and maintenance techniques and equipment must be developed for fibre that transform today's highly skilled task requiring expensive equipment into an everyday task for non-specialists for whom fibre will be only a part of the job. Low-cost fibre splicing, test aids, automatic fibre identification and remote test equipment are all under development.
- Network management computer systems and procedures need to be developed to facilitate the remote hands-off service provision maintenance and fault reporting features inherent in TPON. The order taking and fault reporting (151) functions provided by the CSS computer system need enhancing to accommodate fibre. A small-scale trial to investigate network management issues is currently under way in Manchester.

Management processes need developing and culture changes are needed to facilitate the move from the 'I must have capacity in hand' approach of today's copper pair to the 'just in time' approach of tomorrow's electronics.

4.14 CONCLUSIONS

The deployment of fibre in the access network can help BT to address its problems with the copper network.

Currently, BT has a range of solutions for direct fibres to business customers and, in the longer term, PONs will supplement these for business and residential customers alike. By focusing costs on equipment installed on a just-in-time basis rather than on long-lead-time speculative copper cable capacity, BT can reduce costs and match expenditure more accurately to demand.

That same just-in-time approach coupled with advanced network management and hands-off service provision can allow BT to meet customers' rapid service provision needs.

The self-monitoring ability of electronic systems with the optional facility to duplicate parts of the network, together with a reduction in expedient work causing ‘hands-on’ faults on the copper network, should enable BT to improve its quality of service.

Finally, it has been shown that fibre has an almost infinite capacity for future new broadband services.

REFERENCES

1. Dufour I G: ‘Flexible access systems’, BT Eng J, 7, pp 233-236 (January 1989).
2. Oakley K A, Taylor C G and Stern J R: ‘Passive fibre local loop for telephony with broadband upgrade’, BT Eng J, 7, pp 237-241 (January 1989).
3. Hoppitt C E and Rawson J W D: ‘The United Kingdom trial of fibre in the loop’, BT Eng J, 10, pp 48-58 (April 1991).
4. Fenning S and Rosher P A: ‘Subcarrier multiplexed broadcast video system for the optical field trial at Bishop’s Stortford’, BT Technol J, 8, No 4, pp 26-29 (October 1990).
5. Lisle P and Adams P: ‘Exploiting the copper asset’, BT Eng J, 10, pp 26-33 (April 1991).
6. Howard M H: ‘Blown fibre experience in the local loop’, BT Eng J, 7, pp 242-245 (January 1989).

Part Two

Copper Systems

5

REVIEW OF COPPER-PAIR LOCAL LOOP TRANSMISSION SYSTEMS

P F Adams and J W Cook

5.1 INTRODUCTION

Existing local networks of copper-pair cables provide a medium for digital transmission between subscribers and their local telephone exchange. Transceivers for this medium have undergone a remarkable period of development stimulated by the promise of universal digital communications at rates of 64 kbit/s and above and enabled by rapidly advancing very large scale integrated (VLSI) circuit technology that now allows hitherto unimaginable circuit complexity at low cost, low power consumption and small size. The history of this change and its interaction with standardization processes forms the backdrop of this chapter, which is mainly concerned with the technology of local loop transceivers, especially for the important American National Standards Institute (ANSI) interface standard for integrated services digital network (ISDN) basic rate access. The ANSI standard is outlined and critically appraised and the impact of modern silicon technology trends is assessed on the dichotomy between standardization and functional diversity.

5.2 TECHNOLOGY TRENDS AND LIMITATIONS

Modern high-performance local loop transceivers are a product of rapid advances in silicon technology; to understand why they are designed the way they are requires an appreciation of the trends and limitations of this technology. Three topics are of particular interest — VLSI circuit complexity, analogue circuitry, and analogue-to-digital (A/D) conversion.

5.2.1 VLSI circuit complexity

The complexity of VLSI circuits in terms of gates per chip and chip size has continued to grow over the past three decades and it seems safe to assume that this will continue at least until the end of the century. Such complexity enables the creation of very powerful digital signal processors.

Designing complex circuits is a costly process involving the massive application of computer-aided design (CAD) tools and so is only viable where the development costs can be amortised over enormous product volume. However, two other approaches have emerged to deal with the problem of design costs when the market size in specific applications is less favourable. The first is the general-purpose programmable integrated circuit. Microprocessors are the most obvious examples of this genre. The other approach [1] is to design cells that implement particular functions and then to interconnect these cells using CAD tools to product application-specific circuits. The cells can range from small logic functions to whole subsystems (macrocells).

5.2.2 Analogue circuit trends

In the earlier days of circuit integration there were severe limits on chip size and transistor count. This favoured analogue signal processing implementations which allow more functions per device and do not carry the overheads of A/D and digital-to-analogue (D/A) converters.

The 1970s brought the prevalence of metal oxide semiconductor (MOS) technology which was not well suited to analogue implementations, particularly in its infancy when problems of stability had not been minimized, but was very well suited to densely packed low-power digital circuitry. Therefore, a switch occurred toward digital signal processing (DSP) which was stable, needed no adjustment and could be tested on the basis of identical repeatability.

However, the growth in complexity has reached a point where there is a need to include some inherently analogue subsystems on a chip, especially

A/D and D/A converters. Fortunately fabrication processes have now been developed which allow the inclusion of analogue subsystems. Laser trimming [2] of various on-chip components together with less subtle techniques such as zener-zapping allow critical analogue components to be adjusted on test. Thin-film technology [3] allows precise and extreme-valued linear resistors on a chip surface. Dielectric isolation and other silicon-on-insulator [4] fabrication systems permit high-voltage working. Cell-based design systems [1] permit a wide range of well-characterized analogue components to be available to the designer. Bi-MOS and Bi-CMOS fabrication techniques [5] are well suited to mixed analogue and digital signal processing. Already local loop transceivers [6,7] have been designed in CMOS technology where the whole system — analogue components, digital logic, DSP, and A/D and D/A converters — are implemented on a single chip.

5.2.3 Analogue-to-digital conversion

An A/D converter is the epitome of the unavoidable mixed analogue and digital design situation, and one which has often been avoided, or provided on a separate chip. For modern transceivers, this is often not an acceptable solution. A design methodology that has received significant attention is the exploitation of the trade-off between speed and precision in A/D converter design [8]. Fast over-sampled A/D converters such as sigma delta modulators (SDMs) can be built with comparatively low precision but fast devices. To make the complete A/D converter, an appropriate digital decimating filter is used, reducing the sample rate and increasing the precision. Both parts of this implementation utilize strong points of the technology — speed and compact logic; even so, the resulting A/D converter is rarely competitive, in the silicon area, with a comparable conventional implementation such as a switched capacitor filter and capacitor ladder plus successive approximation register. Oversampling A/D converters have been implemented successfully for transceivers with sampling rates up to 120 ksamples/s [9]. Beyond this, however, the clock speeds required make the technique increasingly unworkable.

5.3 HISTORICAL PERSPECTIVE

The first applications of digital transmission over copper-pair local loops were low data rate baseband modems and limited-range tail circuits [10,11] for digital networks, both operating over four-wire loops. In the 1970s technology limitations and low demand limited transceivers to fairly simple systems [10] in analogue technology. Line codes such as WAL2 [11] were used which have self-equalization properties, good timing content and no energy at zero frequency, all of which obviate the need for complex signal processing. The penalty paid, however, is that the transmitted signal has a high-frequency spectrum that leads to attenuation and crosstalk limitations on system reach.

5.3.1 Burst-mode two-wire duplex transceivers

As the demand for digital transmission increased and especially as the ISDN began to emerge as a possible future service bearer it was realized that duplex working over two wires was essential if demand was to be met at acceptable cost. Of the three possible methods of two-wire duplex working [12], i.e. frequency-division duplex, burst mode and echo cancellation, burst mode was the simplest to realize and so was used for early transceiver designs. BT's integrated digital access pilot service employ such a transceiver operating at 80 kbit/s. However, to obtain duplex working a burst-mode transceiver has to transmit with an instantaneous bit rate of more than twice the user bit rate, leading to more severe problems with attenuation and crosstalk than four-wire systems. The resulting reach limitation confines such systems to short-reach applications such as access to private automatic branch exchanges, or remote multiplexer terminated loops and local loops in city centres.

5.3.2 Early echo-cancelling two-wire duplex transceivers

The limitations of burst-mode systems stimulated the development of echo-cancelling transceivers. Improvements in silicon technology combined with the understanding of adaptive filters arising from their application to speech-band modems allowed the implementation of simple adaptive echo cancellers (ECs). It was found that the self-equalization properties of line codes like WAL2 and biphase also limited the complexity required for ECs. Designs emerged based on analogue implementations [13] and on table look-up techniques [14]. A notable example of the table look-up technology which is commercially available is a single chip 160 kbit/s biphase transceiver [6] developed by Mitel.

5.3.3 Modern echo cancelling transceivers

Transceivers based on self-equalizing codes cannot provide universal coverage of existing local networks. Transceivers using lower-bandwidth codes, which are less susceptible to crosstalk and attenuation limitations, had to be developed. Interest at this point centred on basic-rate access for the ISDN. Theoretical studies [15] based on measured network data suggested that baseband, multi-level codes could give greater than 99% coverage of the UK local network. Similar conclusions were reached in other studies [16] for other networks. To realize this performance required the successful application of a number of novel signal processing techniques which will be described later. Thus there have emerged a variety of designs for 144 kbit/s duplex transmission. These can be subdivided into three classes based on the type of line code employed.

5.3.3.1 Linear line coder designs

A linear line code is one where the spectral characteristics can essentially be obtained by linearly filtering a binary source. Alternate mark inversion (AMI) line code and its variants, e.g. partial response class 4 (PR4), are the most common examples. These required some degree of adaptive equalization especially for networks which include bridged taps. Mixed analogue and DSP implementations of AMI line-code transceivers have been described [17,18]. One interesting scheme [19] used partial burst mode to allow simplification of the EC and timing recovery circuitry.

5.3.3.2 Block code designs

AMI systems for ISDN basic rate access do not quite give sufficient crosstalk-limited reach and so transceivers using block codes with reduced signalling rates were developed by some manufacturers. Two codes are of particular interest: 4B3T and 3B2T (nBmT means that n bits are encoded into m ternary elements).

MMS43, a variant of 4B3T, was chosen for the specification of basic-rate access in Germany and several companies have produced transceiver designs for it. Two single-chip designs [7,20], in particular, exploit an 11-element Barker synchronization word for timing recovery and rely on good linearity in the analogue circuitry. One two-chip design [21] questions the wisdom of using only the synchronization word for timing recovery, and uses nonlinear compensation to relieve the linearity constraints.

SU32, a variant of 3B2T, is the line code used in a two-chip transceiver [22] which also provides the signal-processing functions for implementation of the CCITT Recommendation I.430 S/T interface.

5.3.3.3 2B1Q designs

The very first 2B1Q (two binary one quaternary, i.e. four-level coding) design was that proposed by BT Laboratories (BTL) [23]. This used a very simple frame structure and, by sacrificing some of the excellent performance of 2B1Q, a simple DSP architecture resulted in a potentially compact VLSI circuit design. It employed novel techniques for jitter and nonlinearity compensation. In 1988 an ANSI standard [24] was published for a basic rate access interface that uses 2B1Q.

Unfortunately, as described later, the ANSI standard demands an excessively good performance which has necessitated more complex designs. Examples of transceivers for the ANSI standard which have been published include three-chip [25], two-chip [26,27] and single-chip [28—30] designs including one [29] based on the original BTL design upgraded in line with the ANSI standard.

5.4 ADVANCED TRANSCEIVER TECHNIQUES

An echo-cancelling transceiver can be subdivided into the signal processing blocks shown in Fig. 5.1. The advances in silicon technology enabled complex echo cancellers and equalisers to be used to provide long-reach two-wire echo-cancelling duplex systems. In addition, new timing-recovery techniques and methods of controlling timing jitter, to which echo-cancelling transceivers are particularly sensitive, were required.

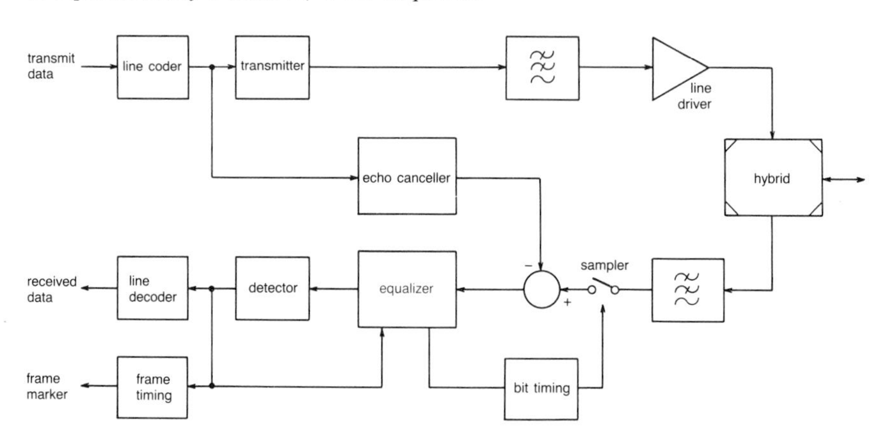

Fig. 5.1 General transceiver structure.

5.4.1 Echo cancellation

The source of echo in a digital transmission system is the transmitted stream of modulated pulses. The echo response can, therefore, be modelled by a data-driven filter. Assuming the echo response to be modelled is a linear function of the current and $N-1$ past data elements then a finite impulse response (FIR) filter can be used (Fig. 5.2). Such a filter can be made adaptive by the use of the least mean squares algorithm. Inherently stable, an FIR filter avoids, in digital implementation, the multiplication of long wordlength numbers because one variable in all the multiplications is a data element. However, terminated copper pairs have slowly decaying echo pulse responses [31] that are significant for many tens of signalling intervals. Therefore, an FIR filter will require a large number of taps to model the echo pulse response adequately. There are two ways round this problem — recursive adaptive tail cancellation and echo tail filtering.

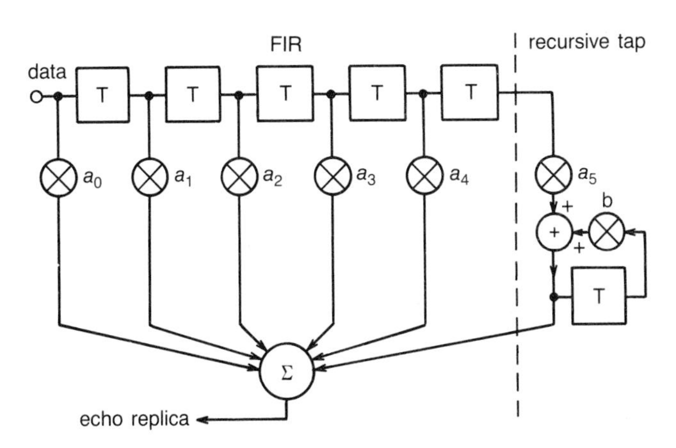

Fig. 5.2 Echo canceller with recursive tap.

5.4.1.1 Recursive tail filtering

It is possible under certain conditions to assume that the slowly decaying tails have an exponential decay function, in which case a single recursive tap added to an FIR EC (Fig. 5.2) can exactly model the tail. Adaptation of such a tap has to be done with care to avoid instability. In a digital implementation the recursive tap requires long-wordlength multiplications to be performed.

5.4.1.2 Echo tail filtering

If a high-pass function is introduced into the overall system frequency response to reduce the low-frequency energy then fewer taps are required in the FIR filter. However, the penalty for this is a degraded performance in the presence of crosstalk.

One class of high-pass functions of interest is $(1-D)/(1-k.D)$, where D is the delay operator and $0<k<1$. Generally as k increases from 0 to 1 the number of EC taps increases rapidly but the crosstalk-limited reach improves steadily. The exact trade-off depends on the line interface design, the target reach of the transceiver and the degree of low-pass filtering in the overall loop response. Overall it is advantageous to have some degree of high-pass filtering.

5.4.1.3 Nonlinearity correction techniques

A further problem that can arise in using a linear EC is that nonlinearities in the transceiver's transmitter and line-interfacing circuitry can degrade performance. Usually the echo is required to be cancelled to more than 20 dB below the received signal which itself may be up to 40 dB below the transmitted signal on a long loop. Therefore, even small nonlinearities can cause problems. The echo to be modelled by the echo canceller can be expressed as a general function of the past N data elements which can be expanded as a Volterra series. Nonlinearity correction involves either the addition of extra taps [32] that individually model each term in the Volterra expansion, or the application of table look-up techniques [33]. The former is fine if the significant terms of the Volterra expansion are known; the latter requires no specific knowledge of the nonlinearity, only its location in an otherwise linear channel.

The table look-up technique assumes that the echo path can be modelled by a memoryless nonlinear function preceded by a linear function of N^* data elements and followed by a further linear function, as shown in Fig. 5.3. Such a model can be mapped into a structure obtained by summing the outputs of N' random access memories (RAMs) which are each addressed by sets of N^* consecutive data elements as shown in Fig. 5.4. Used in parallel with a linear adaptive filter such an adaptive nonlinearity modelling structure can be very effective [33]. If the total significant echo span is short enough then this structure can be reduced to a single look-up table by putting $N^*=N$ and $N'=1$. The echo span of baseband line codes rules out this simplification, especially for M-level codes, where the size of the memory is M^{N}. However,

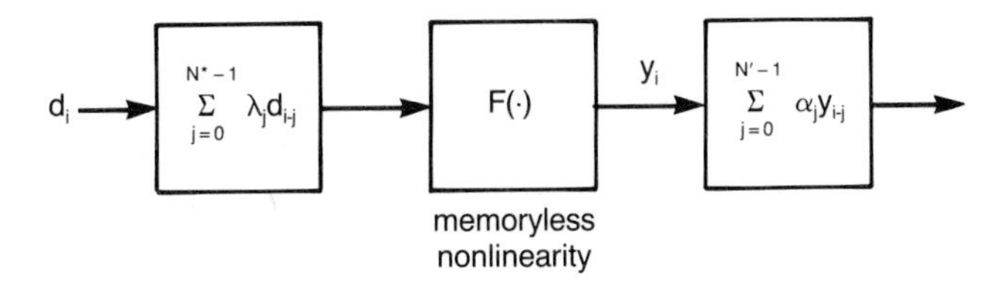

Fig. 5.3 Echo path model.

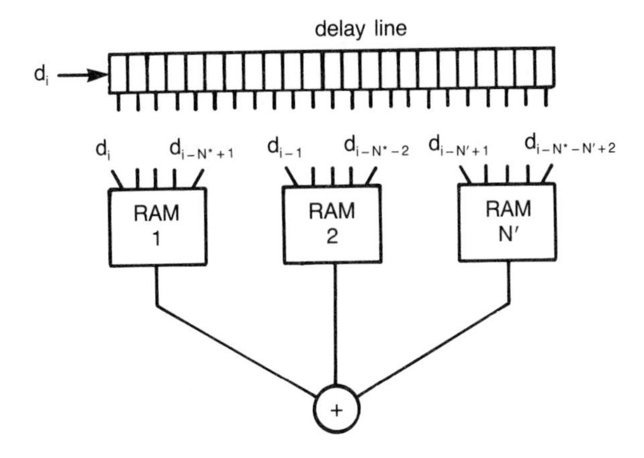

Fig. 5.4 Table look-up nonlinearity modelling structure.

they have been used [6] successfully in transceivers with self-equalizing line codes which create a short significant echo response. When they can be used, table look-up ECs are ideally suited to VLSI circuit implementation, being predominantly RAM and a single adder for memory adaptation.

5.4.2 Equalization

The most general form of equalisation contemplated so far for local loop transceivers, which is shown in Fig. 5.5, consists of a decision feedback equalizer (DFE) preceded by a pre-cursor equalizer (PCE) and a linear tail equalizer (LTE). All the equalizers are adaptive and attempt to drive the error signal to zero. The received pulses are sampled at, or near, the pulse peak and the PCE reduces pre-cursor inter-symbol interference (ISI), while the DFE cancels post-cursor ISI and the LTE reduces the pulse tail outside the span of the DFE. The DFE is identical in structure to the EC. The PCE and LTE, however, involve multiplications of tap values by received signal samples and so are more complex to implement digitally.

With some loss of performance it is possible to sample the received pulses just prior to their peak such that the pre-cursor ISI is negligible and there is no need for a PCE. As in the case of echo tails, judicious high-pass filtering can be used to reduce the pulse tail to the point where an adaptive LTE is not needed.

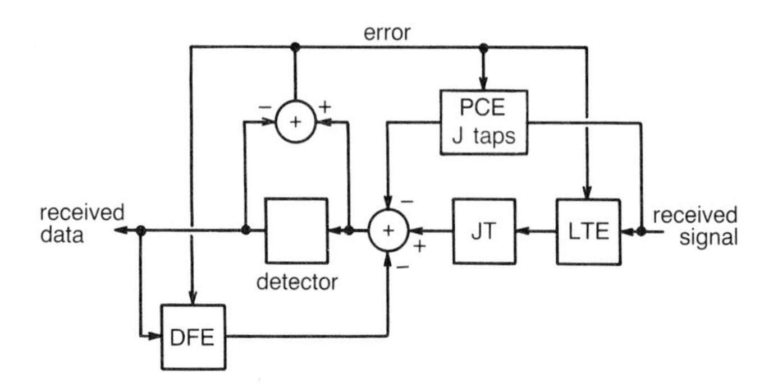

Fig. 5.5 General adaptive equalization for local loop transceivers.

5.4.3 Timing recovery

Commonly, timing recovery involves exploitation of a cyclostationary property of the received signal. This allows a timing phase to be established and maintained without prior equalization or data detection. However, it does not necessarily result in adequately small pre-cursor ISI and so, often, necessitates a PCE to obtain a good performance over all loops. One example [34], where the cyclostationary property is obtained by use of a block frame synchronization work (SW) repeated at 1 ms intervals, is that defined in the German specification for ISDN basic rate access. The SW is an 11 element Barker word,

$$+2,\ +2,\ +2,\ -2,\ -2,\ -2,\ +2,\ -2,\ -2,\ +2,\ -2,$$

which has an autocorrelation function peak at zero shift of 44 and all non-zero shift terms of modulus less than or equal to 4. This property means that when the received signal is filtered appropriately the output of the filter always produces a zero crossing at a specific time coincident with the end of the synchronization word and near the pulse peak of the channel response. Such a signal can be used to drive a phase-locked loop (PLL).

An optimum timing phase that can avoid the use of PCE can be obtained by the use of a decision-directed (DD) timing scheme. DD schemes [35] are based on the premise that a specific function of both the received signal and the correctly detected data elements gives a monotonic output that passes through zero at the desired sampling phase. They do require joint adaptation with the equalizer, but this can be advantageous as it can help to prevent DFEs locking up during training.

Timing-recovery schemes which do not rely on a block SW have the advantage that all the received signal is used for timing phase estimation. Therefore they are, for a given tracking rate, less sensitive to noise and hence give better jitter performance.

5.4.4 Jitter control

ECs are very sensitive to clock jitter and so careful design of the timing-recovery scheme is necessary to prevent it from degrading performance. Timing-recovery schemes that employ analogue PLLs have to be designed with high Q [36] if they are not to impair EC performance significantly.

A digital PLL makes sudden discrete phase adjustments which, if the ratio of its clock rate to the signalling rate is not high enough, will degrade the EC performance unacceptably. Three methods have been used to prevent this — zero echo gradient phase adjustment, frame word synchronous phase adjustment and jitter compensation. For the first method to work the signal transmitted by the local transmitter must incorporate a period when the signal is at constant amplitude for long enough that a change in the receive sampling phase does not cause any change in the echo. A particular example of this is the 'echo pong' transceiver described in Mogavero *et al* [19] where the transmitted signal is set to zero periodically.

For the frame word synchronous phase adjustment technique phase adjustment is done during the reception of the frame synchronization word. As this word is not customer's data, corruption of it by uncancelled echo is allowable. This technique does, however, force the use of a stored replica of the synchronization word to prevent error propagation if a DFE is used. Also, like the zero echo gradient technique, it is not effective in the LT transceiver because of the non-transient nature of the change in the echo response.

Jitter compensation, as shown in Fig. 5.6, places no constraints on the design of signals. It also allows digital PLL timing phase tracking in the LT transceiver. It was first described in publications [37,38] from BTL in 1985. A less versatile version was subsequently described by Messerschmitt [39]. The BTL technique was subsequently re-described almost exactly in Kanemasa *et al* [40], with the added extension of the use of the sign update least means square adaptation algorithm. Essentially, jitter compensation involves

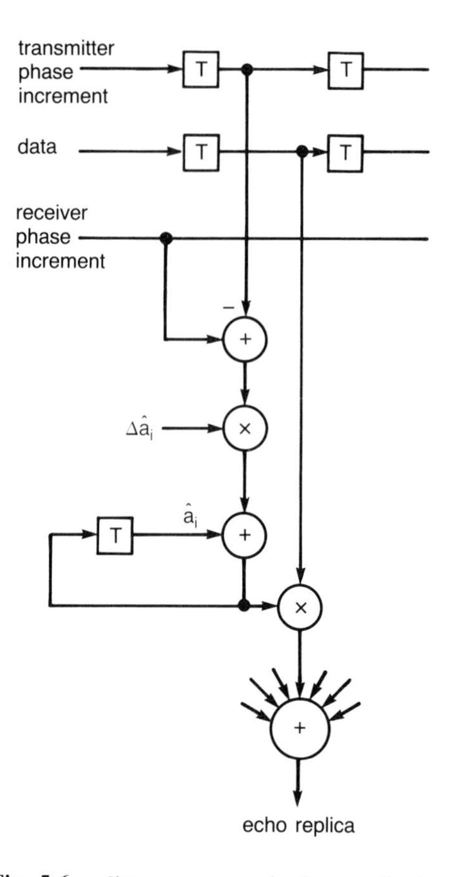

Fig. 5.6 Jitter-compensated echo canceller tap.

adaptively generating a gradient estimate $\Delta\hat{a}_n$ for the *n*th tap of the FIR filter which is used to correct the tap under control of the transmit and receive phase increments as illustrated in Fig. 5.6.

5.5 STANDARDIZATION

In the UK and many other countries the network boundary of the ISDN from a regulatory point of view is on the subscriber's side of the network terminating equipment (NT1). Beyond the end of the network the subscriber can procure equipment competitively which will work to a standard interface, i.e. CCITT Recommendation I.430.

Such a regulatory environment is highly sensible from a network management and maintenance point of view because it allows network continuity and performance to be monitored and in-built diagnostics. The effect on local loop equipment procurement policy is that only the functionality and performance of transmission equipment within a defined local loop environment needs to be specified.

In the USA a different regulatory regime applies for the ISDN. The network boundary is defined as the end of the pair of wires entering the subscriber's premises. It is, therefore, necessary to define an interworking interface for the end of the network. Such an interface requires very rigorous definition of the electrical signals on it if interworking between transceivers from different manufacturers is to be achieved. This is not an easy task especially as the signals from the network can be severely distorted in a loop-dependent way.

Maintenance and loop diagnostic features can be included by defining the meaning of bits in a maintenance channel added to the raw ISDN basic rate-access channels. However, the need to have a frozen standard, or at least one that evolves only very slowly, means that upgrading to improved maintenance features is very difficult and any changes must be compatible with existing equipment.

The US basic rate access interface standard is a primary driver towards transceiver standardization. The defined functionality and performance requirements of the US standard mean that chip manufacturers can address a large stable market, a prerequisite for the development of complex integrated circuits if a low unit cost is to be achieved.

5.6 US ANSI STANDARD FOR BASIC RATE ACCESS

The US ANSI standard [24] for the interface on the network side of the NT1 has emerged as an important influence on the development of local loop transceivers and so will now be described and assessed in some detail.

The standard attempts to define completely the signals that appear at the subscriber's end of a local loop from both the line termination (LT) and the NT1. Central to this definition is the choice of line code which has a profound effect on the digital transmission performance and the complexity of implementation required to realize that performance.

The standard specifies quaternary pulse amplitude modulation which was first used for local loop transmission by BTL in 1983; the name 2B1Q was proposed by BTL in 1986 during the debate on the choice of the line code for the ANSI standard.

5.6.1 Technical justification for 2B1Q

Two issues dominate the choice of line code — near-end crosstalk (NEXT) limited reach and implementation complexity.

The NEXT limited reach of a transmission system is a function of:

- the number of levels in the line code and its power spectral density;
- the modulation rate;
- the system transmit and receive filtering;
- the type of equalization;
- the data-detection method.

These factors are not entirely independent of each other and the performance of line codes can vary as each of the other factors is changed. Comparisons based on a common shape of transmitter and receiver filtering scaled by the modulation rate give results typified in Fig. 5.7 for a transceiver using only a DFE.

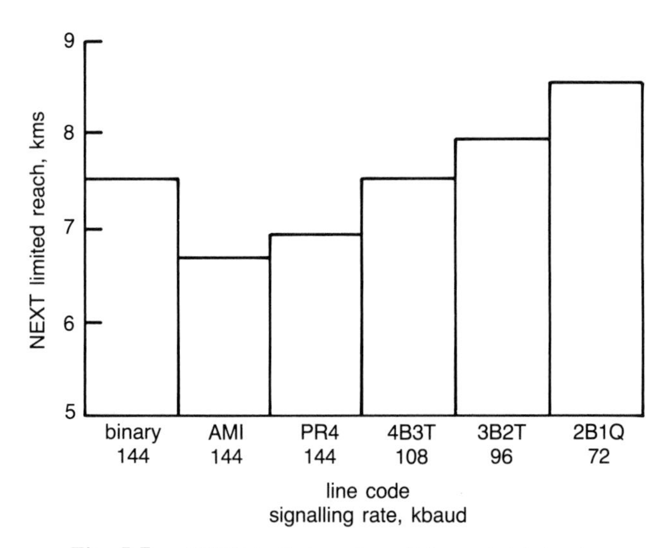

Fig. 5.7 NEXT limited reach on 0.5 mm twisted pair.

Clearly multi-level, low redundancy block line codes give performance advantages, a result confirmed in Adams and Cox [41]. However, as linear line codes such as AMI can give almost sufficient NEXT limited reach with simpler transceiver designs, is the performance advantage of block line codes worth the increased complexity? Service providers are keen to have long reach so that special engineering of local loops can be avoided, but only at acceptable cost. 2B1Q resolves the argument by combining the advantages of both linear and block line codes. Practical measurement of transceiver performance indicates that 2B1Q has a slightly superior performance to ternary block codes and a significant gain over linear line codes. Assessed on replicas of the 15 test loops chosen by Bell Communications Research (Bellcore) to be representative of the longest 15% of US loops a simple 2B1Q transceiver designed by BTL produced the results shown in Table 5.1 for both forward (F) and reversed (R) directions of each loop. The crosstalk power sum attenuation was 57 dB at 80 kHz with a slope of 4.5 dB per octave. Also presented are some simplex simulation results for a more complex transceiver with a DFE, LTE and PCE, each with a large number of taps, which are close to the best achievable with such an equalization strategy. Using data specified by Bellcore [42] for a 6 dB NEXT margin these results predict loop coverages of 97.8% and 99.9% respectively.

Table 5.1 2B1Q next margin (dB) for a 10^{-7} error rate over the Bellcore loop set.

Bellcore loop no	*1986 BT Labs measurement results*		*BT Labs simple simulation results*
	F	R	
1	−4	−5	4
2	0	−4	8
3	2	6	11
4	4	3	9
5	8	5	11
6	7	5	11
7	7	7	12
8	5	4	13
9	11	8	14
10	9	8	14
11	12	10	15
12	10	9	14
13	13	10	16
14	10	11	15
15	10	11	15

As practical measurement demonstrates a small performance advantage for 2B1Q, transceiver implementation complexity issues become an important factor. As will be illustrated later the evidence [43] was that 2B1Q transceivers can be implemented with less chip area than other transceivers of equivalent performance.

5.6.2 Signalling rate and frame synchronization

For 2B1Q to encode the 144 kbit/s of ISDN basic rate access the modulation rate must be at least 72 kbaud. However, to allow the identification of the channels that make up the total 144 kbit/s, a frame structure is necessary which incorporates frame alignment information. In addition, an overhead capacity is required for maintenance and control purposes. The standard specifies a modulation rate of 80 kbaud, giving 4 kbit/s overhead capacity and 12 kbit/s for frame synchronization.

A frame structure is defined based on the use of a block SW of nine quaternary elements (quats) repeated every 1.5 ms. This allows the use of frame word synchronized timing adjustments if required. The 9-quat word

$$+3,\ +3,\ -3,\ -3,\ -3,\ +3,\ -3,\ +3,\ +3$$

is more efficient in the lower baud rate 2B1Q standard than the 11 element word of the German specification but has similar properties, so allowing joint bit and frame timing recovery. However, the shorter duration in baud intervals and the lower repetition rate mean that it is less effective for these synchronization techniques.

5.6.3 Transmit signal specification

The standard defines key features of the transmitted line signal tightly enough to avoid undue performance variation when transceivers from different manufacturers have to interwork, but still allows manufacture with reasonable component tolerancing. These features are transmit signal power, power spectrum, peak pulse amplitude, pulse shape, maximum nonlinear distortion and jitter.

5.6.3.1 Transmit signal power, power spectrum and nonlinearity

The standard specifies a transmit power of 13.5 dBm ±0.5 dBm measured across a 135 Ω resistive load. This signal power level was judged to be a good compromise between adequate impulse noise tolerance and excessive crosstalk interference into other services. A reduced power of 11.6 dBm ±0.5 dBm is allowed up to 1992. The reason for this is that it was believed to be difficult initially to generate 13.5 dBm with acceptable linearity using 5 V CMOS

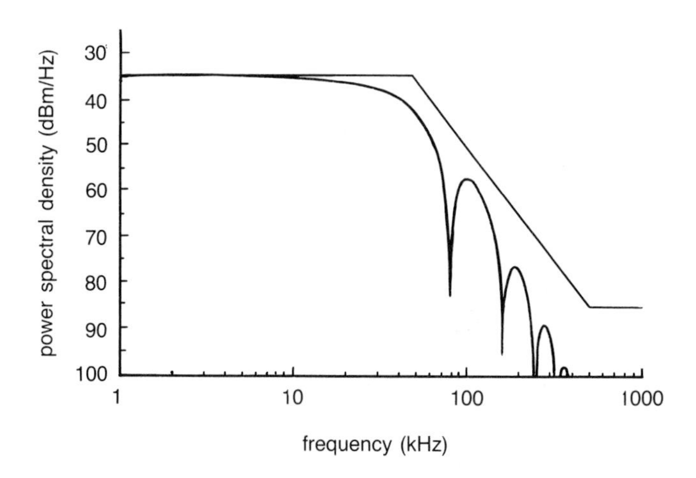

Fig. 5.8 Power density spectral mask and response.

technology. A penalty of this compromise is to cause a 2 dB degradation of NEXT performance when both co-exit.

The standard imposes the power spectrum mask shown in Fig. 5.8 to prevent unnecessary interference into other services. Figure 5.8 also shows a typical power spectrum.

Transmitter nonlinearity can cause performance degradation at the far-end receiver if the equalizer can only model a linear response. The standard specifies the total nonlinear power at the transmitter relative to the signal power when measured across a 135 Ω load to be −36 dB or less. This gives an acceptable signal to nonlinearity noise ratio at the far-end receiver allowing for the increase in ratio of signal power to pulse amplitude at the end of a long loop.

5.6.3.2 Pulse mask and peak voltage

The pulse mask shown in Fig. 5.9 is specified to constrain the transmitted signal adequately to avoid excessive performance variation whilst still allowing reasonable component tolerancing. The pulse mask specifies a pulse peak of 2.5 V (2.0 V until 1992) for a +3 quat. Unfortunately, a wide variation in the pulse tail shape within the mask is possible which can result in performance variations when various designs of transceiver are required to interwork. Figure 5.9 also shows a typical pulse response that meets the mask.

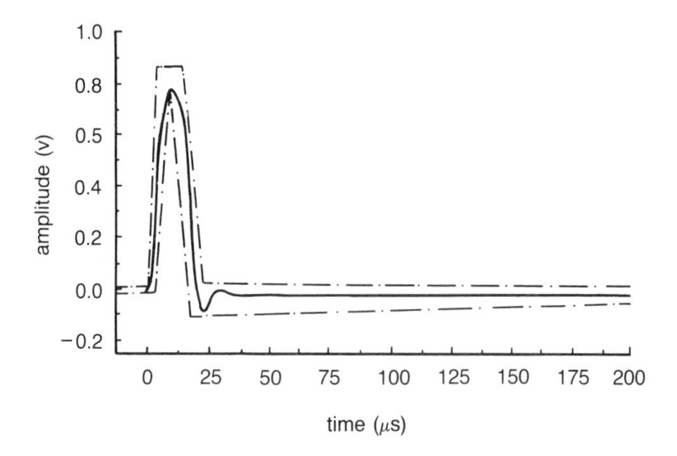

Fig. 5.9 Pulse mask and +1 quat time response.

5.6.3.3 Frequency offset and jitter

The pulses received at the NT1 should arrive every 12.5 μs. However, frequency offset on the network clock to which the LT is slaved causes the pulses to arrive more or less frequently. In the US the network clock tolerance is specified as 5 ppm. The network clock can also suffer from jitter. The NT1 must track both the frequency offset and low-frequency jitter. High-frequency jitter cannot be tracked but normally it is small enough not to cause any problems.

The standard assumes that the jitter on the network clock is no worse than sinusoidal jitter of a frequency and amplitude constrained by a mask. Above 19 Hz the jitter has a maximum peak-to-peak value of 0.1 μs. Below 19 Hz the maximum peak-to-peak jitter increases at 20 dB/decade to a maximum of 3.75 μs at 0.5 Hz and below. This degree of low-frequency jitter is approximately equivalent to having to track an additional 6 ppm frequency offset. The NT1 must track frequency offset relative to its own free-running clock frequency — typically, up to 50 ppm offset from nominal. In this case the NT1 clock recovery circuit must be capable of tracking approximately 61 ppm worst-case relative frequency offset.

The pulses transmitted by the NT1 will be timed from the NT1's 80 kHz clock. This tracks the LT clock but will contain some jitter caused by imperfect timing recovery. The standard limits this jitter by both r.m.s. and peak constraints.

As the effect at the LT receiver of jitter on the received pulses is to generate a noise source, which is the sum of a number of uncancelled ISI gradient terms, r.m.s. constraints are appropriate. The power of the noise source is related to r.m.s. jitter. However, r.m.s. constraints alone limit only the average effect of the jitter. To limit short-term peak jitter from causing temporarily large error rates and possibly receiver instability peak-to-peak constraints are also applied. However, the standard does not specify an observation period and so such constraints are statistically ambiguous.

The jitter constraints are defined for two frequency bands — a high-frequency band (>80 Hz) to prevent performance degradation in the LT receiver arising from untrackable jitter, with constraints of 0.125 μs r.m.s. and 0.5 μs peak to peak, and a lower frequency band (>1 Hz, <40 Hz) to prevent excessive jitter relative to the LT clock, with constraints of 0.01875 μs r.m.s. and 0.625 μs peak to peak. In addition, excessive wander is prevented by an overall limit on the relative phase difference between the NT1 and LT clocks of 1.25 μs peak after subtraction of the mean phase difference.

5.6.4 Start-up procedure

Each transceiver requires that a defined sequence of signals occurs across the interface during a start-up phase to train the adaptive filters and establish stable timing phases. The standard allows 15 s for a cold start, and 300 ms for a warm start to establish normal operation. The apportionment of this time between the LT and NT1 transceivers is illustrated in Fig. 5.10 which shows the individual segments of the training signals. A and C are controlled by the NT1 transceiver and B and D by the LT transceiver. For a cold start $A+C<5$ s and $B+D<10$ s and for a warm start both must be <150 ms. The time apportioned can be used differently in different transceiver designs.

The original BT Laboratories 2B1Q transceiver design used binary signal segments during start-up to allow very rapid equalizer training from reset. However, a simple cold-start procedure (activation/deactivation of the interface being optional) was of prime concern in the standard and so segments using only quaternary signals were defined. The absence of binary signal segments means that the warm-start specification can be met only by prior storage of adaptive filter coefficients and/or sampling phase at the LT. These would be stored after a cold start and subsequent deactivation. However, reliable warm start-up may not be possible where large temperature variations give rise to significant changes in cable characteristics between deactivation and warm start.

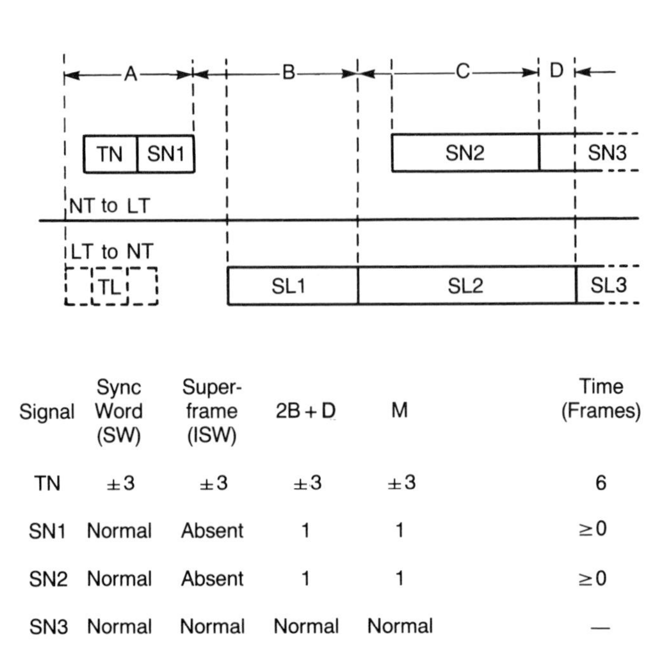

Signal	Sync Word (SW)	Super-frame (ISW)	2B + D	M	Time (Frames)
TN	±3	±3	±3	±3	6
SN1	Normal	Absent	1	1	≥0
SN2	Normal	Absent	1	1	≥0
SN3	Normal	Normal	Normal	Normal	—
TL	±3	±3	±3	±3	2
SL1	Normal	Absent	1	1	≥0
SL2	Normal	Normal	0	Normal	≥0
SL3	Normal	Normal	Normal	Normal	—

Fig. 5.10 Start-up signals.

5.6.5 NEXT performance specification

The standard primarily defines the signal on the interface transmitted by the NT1. It is assumed that the LT transmits a similarly defined signal (apart from its jitter specification). However, the signal arriving at the interface has its electrical characteristics drastically modified by the copper pair response which varies from loop to loop, and so is very difficult to define. Electrical definition of this signal is avoided by specifying test loops over which satisfactory NEXT error rate performance is required. The standard requires operation over Bellcore's test loops 4-15 with 6 dB margin for a 1 in 10^7 error rate against a standard NEXT noise source. The NEXT noise source is defined as equivalent to the sum of two non-return-to-zero pulse streams, one at 80 kbaud, the other at 160 kbaud. Although this is a signal that violates the pulse power spectrum mask, the intention was to produce a wide power spectrum that forces a sharp cut-off above 50 kHz at the input to the transceivers to ensure that interference from wider bandwidth signals was adequately rejected.

5.7 IMPLEMENTATION OF THE STANDARD

All the advanced transceiver techniques described earlier can be applied with advantage to the implementation of transceivers for the ANSI standard. Referring to Fig. 5.1 it is assumed that the EC, equalizer and quantizer are realized by DSP. Analogue realizations have been advocated [44] because they lead to a reduced chip area, but as previously discussed there are stronger reasons for using DSP. The timing-recovery function, receive filter and the transmitter are all amenable to a variety of mixed analogue and digital implementations.

5.7.1 DSP implementation

The EC and equalizer are the most significant DSP functions. It is important, therefore, that their complexity is minimized for the smallest chip size and power consumption. One of the advantages of multilevel codes is that the reduced signalling rate required reduces the number of taps needed in the adaptive filters and allows more time for the signal processing to be performed. This is illustrated in Table 5.2 for a number of line codes and a particular transceiver architecture for equal NEXT limited reach. A reduction in the number of taps equates to a reduction in the amount of storage required; a reduction in the processing rate equates to reduced chip area for processing circuitry assuming that a processor architecture is used to exploit the reduced rate. Clearly, 2B1Q gives the significant advantage of reduced complexity. However, this advantage does raise the issue of multiplications by quaternary data.

Table 5.2 EC and DFE complexity and processing rate for different line codes.

Line code	*Baud rate*	*Total number of taps*	*Processing rate (taps/μs)*
AMI	160	80	12.6
MDB	160	98	15.6
MMS43	120	89	10.6
SU32	108	65	7.1
2B1Q	80	53	4.2

Quaternary multiplications in the EC and DFE appear to require more processing than a binary multiplication, i.e. a conditional addition operation is required to multiply by 1 or 3. Algorithm modification and re-ordering of processing, however, can reduce this extra processing to a negligible amount.

Adaptation algorithms for the EC, for example, can be modified to use only the sign of the data with only a moderate increase in convergence time [31].The tap multiplications can be simplified by accumulating the tap values to be multiplied by 3 and by 1 (after multiplication by the sign of the appropriate data element) separately and multiplying the partial sums by 3 and 1 respectively before adding them together. These two modifications result in the number of addition/subtraction operations only increasing by three over a binary echo canceller.

5.7.2 Analogue implementation issues

Transceiver performance is ultimately limited by its analogue circuitry [45]. The transmitter must provide accurate pulse amplitudes and shapes, and line drivers must have low-distortion. The A/D converter must have sufficient signal-to-noise ratio to achieve the reach required in the standard. These objectives are not particularly more difficult to achieve with quaternary signals than with binary or ternary signals.

5.7.2.1 Transmitter implementation

It is argued that a quaternary signal will have potentially 16 different pulse shapes if the four levels are all offset from their ideal values and the transition times between adjacent elements are different for any pair of elements, compared with nine and four for ternary and binary signals respectively and, therefore, a quaternary signal is more difficult to generate with a given degree of linearity than a binary or ternary signal. In practice the transition variations can be avoided if adjacent pairs of pulses are generated by separate pulse generators. It is possible to view the generation of a quaternary signal as the sum of two binary signals with their amplitudes in the ratio 2:1. As forming components that have an accurate ratio in silicon technology is straightforward, it is not particularly difficult to construct a quaternary transmitter.

The linearity requirements are dictated by the required level of echo cancellation relative to the received signal level. The possible use of nonlinearity correction affects the requirements, subject to the transmit signal linearity specification of the standard.

5.7.2.2 Line driver

The design of the line driver is affected by the peak current it has to deliver to the line transformer and the tolerable level of nonlinear distortion. The peak current is proportional to the peak-to-r.m.s. ratio of the line code and the level of nonlinear distortion proportional to the pulse attenuation suffered by the line code. Taking these two factors together 2B1Q appears to require a slightly lower performance from the line driver than binary or ternary line codes because its reduced pulse attenuation outweighs its greater peak-to-r.m.s. ratio.

5.7.2.3 A/D converter

The A/D converter is required to have a better than 70 dB signal-to-noise ratio over a minimum of 40 kHz bandwidth. This signal-to-noise ratio is not significantly different from other line codes and the reduced bandwidth of 2B1Q can make the design simpler.

The oversampled DSM A/D converter described earlier is an attractive technique. The digital decimating filter required to filter the out-of-band quantization noise can also provide accurate pulse shaping and fixed equalization in such a way as to minimize the effects of dispersion on the complexity of any nonlinearity and jitter compensators in the echo canceller.

5.8 NON-BASIC RATE TRANSCEIVERS

The heavy focus on the basic-rate application and in particular the ANSI standard has produced several designs most of which implement the ANSI 2B1Q standard. Currently, however, most of the digital local loop systems installed are for digital private circuit access and support only a single 64 kbit/s channel. The degree to which service integration on the ISDN will come about is not at all clear. Different services often require different transmission and maintenance functionality which is difficult to contain in a single standard without making the standard overly complex.

Technology is providing viable alternatives to the custom-designed VLSI circuit approach, such as the macrocell and microprocessor approaches which may allow some degree of versatility in functional specification, albeit at higher cost. The microprocessor approach also allows retrospective design changes to be made by changes in firmware.

Resolution of the US ANSI standard and pressures to make better use of the installed base of copper pairs has raised the question of higher-rate transmission. Theoretical studies [15,16] suggest that 85% coverage of the UK network may be possible at 384 kbit/s and 70% at 768 kbit/s. The use of complex modulation and detection schemes has been proposed [46,47] to further increase coverage, but realization difficulties may preclude the more complex schemes.

Initial studies [48] suggest that the NEXT-limited reach of higher-rate systems will be less affected by the choice of line code provided the signalling rate is less than about 500-600 kHz and is used with an efficient line code.

The reduced signalling interval required for higher-rate systems will mean that implementation will require either precision circuitry for analogue signal processing, or very efficient digital processing algorithms.

Recent developments [49] have shown that simple processing algorithms can be contained on two or three DSP microprocessor devices to produce transceivers that produce the loop coverages predicted above.

5.9 CONCLUSIONS

Copper-pair local-loop transceiver technology has now matured to the point where echo-cancelling transceivers giving nearly 100% network coverage at rates up to 144 kbit/s are emerging as single VLSI circuits of potentially low-cost, small size and low power consumption. The focus of this technology is the ISDN basic rate access, especially as defined by the US ANSI standard. However, evolving technology and service and functional diversification may mean a variety of local loop transceivers will result, transporting different user rates.

REFERENCES

1. Bienstman L B: 'The use of analogue and digital cell libraries in BIMOS', 6th Int Conf on Custom and Semicustom ICs, pp 54/1-54/6 (November 1986).

2. Swenson E J: 'Present and future applications for laser processing of hybrids and semiconductors', Proc SPIE Int Soc Optical Engineering, 527, pp 45-50 (USA) (January 1985).

3. Godfrey R: 'Thin film circuits — a fast improving technology', New Electron, 19, No 11, pp 28-29 (GB) (May 1986).

4. Celler G K: 'Si-on-insulator for ULSI applications', Proc of the 1st Int Symp on Ultra Large Integration, pp 696-711 (May 1987).

5. Henley S: 'BiCMOS shrinks below a micron', New Electron, 21, No 4, p 35 (UK) (April 1988).

6. Colbeck R P and Gillingham P B: 'A 160 kbit/s digital subscriber loop transceiver with memory compensation echo canceller', IEEE JSSC, SC-21, No 1, pp 65-72 (February 1986).

7. Sallaerts D et al: 'A single-chip U-interface transceiver for ISDN', IEEE JSSC, SC-22, No 6, pp 1011-1021 (December 1987).

8. Candy J: 'A use of limit cycle oscillations to obtain robust analog-to-digital converters', IEEE Trans Com, COM-22, pp 298-305 (March 1974).

9. Roessler B, Wolter E and Sporer H: 'CMOS analogue front end of a transceiver with digital echo cancellation for ISDN', IEEE Custom Integrated Circuits Conference, pp 457-460 (1987).

10. Bigg R W: 'Modems for 48 kbit/s data transmission', POEEJ, 2, Part 2, pp 116-122 (July 1971).

11. Boulter R A: 'A 60 kbit/s data modem for use over physical pairs', Proc Zurich Seminar on Digital Communication, p H3/6 (1974).

12. Griffiths J M: 'ISDN explained', Wiley (1990).

13. Adams P F, Glen P J and Woolhouse S P: 'Echo cancellation applied to WAL2 digital transmission in the local network', IEE Conf publication 193, Telecommunication Transmission in the Digital Era, pp 201-204 (1981).

14. Justnes B O: 'A transmission module for the digital subscriber loop', Proc Communications '80, pp 73-76, Birmingham, England (1980).

15. Cox S A and Adams P F: 'An analysis of digital transmission techniques for the local network', BT Technol J, 3, No 3, pp 73-85 (July 1985).

16. Joshi V and Falconer D D: 'Channel capacity bounds for the digital subscriber loop', IEEE Pacific Rim Conf on Communications, Computers and DSP, pp 202-204 (June 1987).

17. Arnon E, Chomik W and Elder M: 'A transmission system for ISDN loops', ICC, Toronto, Canada, pp 7.5.1-7.5.9 (1986).

18. Blake R B et al: 'An ISDN 2B + D basic access transmission system', Proc ISSLS, pp 256-260 (1986).

19. Mogavero C, Nervo G and Paschetta G: 'Mixed recursive echo canceller (MREC)', Conf Rec IEEE Globecom, 1, pp 44-48 (December 1986).

20. Sailer H, Schenk H and Schmid E: 'A VLSI transceiver for the ISDN customer access', ICC '85, pp 45.4.1-4.4 (1985).

21. Wouda K et al: 'Towards a single-chip ISDN transmission unit', Proc ISSLS '86, pp 250-255 (1986).

22. ALT144/DSP144, 'SU32 ISDN U-interface transceiver', STC plc data sheet (Jan 1988).

23. Adams P F et al: 'A long reach digital subscriber loop transceiver', Globecom, Houston, USA, pp 39-43 (1987).

24. ANSI standard T1.601—1988: 'Integrated services digital network (ISDN) — basic rate access interface for use on metallic loops for application on the network side of the NT1 (layer 1 specification)' (Sept 1988).

25. Takahashi Y et al: 'An ISDN echo cancelling transceiver chip for 2B1Q coded U-interface', ISSCC, New York, USA (1989).

26. Khorromahedi et al: 'An ANSI standard ISDN transceiver chip set', ISSCC, New York, USA (1989).

27. Koch R, Niggebaum R and Vogel D: '2B1Q transceiver for ISDN subscriber loop', ISSCC, 360, pp 260-261, New York, USA (1989).

28. Arnon E, Chomik W and Aly S: 'Performance of a 2B1Q transmission system for ISDN basic access', ISSLS '88, Boston USA, pp 2.2.1-2.2.5 (1988).

29. Colbeck R P: 'A single-chip 2B1Q u-interface transceiver', ISSCC, 24, No 6, pp 1614-1624, New York, USA (1989).

30. Batruni R et al: 'Mixed digital/analogue signal processing for a single chip 2B1Q interface transceiver', ISSCC, 25, No 6, pp 1414-1425 (1990).

31. Adams P F, Cox S A and Glen P J: 'Long reach duplex transmission systems for the ISDN', BT Technol J, 2, No 3, pp 35-42 (April 1984).

32. Aggazi O, Hodges D A and Messerchmitt D G: 'Non-linear echo cancellation of data signals', IEEE Trans. COM-20, pp 2421-2433 (Nov 1982).

33. Cowan C F N and Adams P F: 'Non-linear system modelling: concept and application', Proc ICASSP, '84, pp 4561-4564 (1984).

34. Szechenyi K, Zapp F and Sallaerts D: 'Integrated full-digital u-interface circuit for ISDN subscriber loops', IEEE JSAC, SAC-4, No 8, pp 1337-1349 (Nov 1986).

35. Mueller K H and Muller M: 'Timing recovery in digital synchronous data receivers', IEEE Trans COM-24, No 5, pp 516-531 (May 1976).

36. Brophy S G and Falconer D D: 'Investigation of synchronisation parameters in a digital subscriber loop transmission system', IEEE JSAC, SAC-4, No 8, pp 1312-1316 (Nov 1986).

37. Cox S A: 'Clock sensitivity reduction in echo cancellation', Electronic letters, 21, No 14, pp 585-586 (July 1985).

38. Carpenter R B P, Cox S A and Adams P F: 'Jitter compensation in echo cancellers', IASTED Int Sym on Applied Sig Proc and Dig Filtering, Paris, France (1985).

39. Messerschmitt D G: 'Asynchronous and timing jitter insensitive data echo cancellation', IEEE Trans, COM-34, No 12, pp 1209-1217 (Dec 1986).

40. Kanemasa A et al: 'Compensation for the residual echo increase due to timing clock phase jump', Globecom '87, 3, pp 1971-1975 (Nov 1987).

41. Adams P F and Cox S A: 'The 2B1Q line code: genesis and transceiver implementation', ISSLS '88, Boston, USA, pp 19-23 (1988).

42. Bellcore, T1D1 document T1D1.3/86-095 (1986).

43. Adams P F: '2B1Q — A standard for the 'U' interface in the USA', Midcon, pp 453-456 (Sept 1987).

44. Agazzi O, Hodges D A and Messerschmitt D G: 'Large-scale integration of hybrid method digital subscriber loops', IEEE Trans, COM-30, pp 2095-2108, (Sept 1982).

45. Adams P F and Colbeck R P: 'Considerations in the implementation of 2B1Q transceivers', Proc ISCAS, 1, pp 861-864, Espoo, Finland (1988).

46. Lechleider J W: 'Feasibility study of very high bit rate digital subscriber lines', Bellcore contribution to T1E1.4, T1E1/88—012 (March 1988).

47. Lin D W: 'Wideband digital subscriber access with multidimensional block modulation and decision feedback equalisation', Globecom '88, pp 25.5.1-25.5.5 (Nov 1988).

48. Hunt R G et al: 'The potential for high rate digital subscriber loops', ICC '89, 1, pp 520-525, Boston, USA (1989).

49. Young G and Cole N G: 'Design issues for early high bit rate digital subscriber lines', Globecom '90, pp 1177-1182 (1990).

6

IMPROVED TECHNIQUES FOR ACCESS NETWORK FAULT LOCATION

J Trigger

6.1 INTRODUCTION

The rapid modernization of the trunk and junction networks is resulting in significantly improved levels of communications quality and network maintainability and these must be carried through to the local network if the customer is to receive the improvement in quality of service this promises. Despite the increasing use of optical fibre and radio systems it is the metallic pair network that warrants most attention since, at least for the foreseeable future, it will remain the transmission medium for most customers.

This expected longevity of the metallic network brings with it some problems.

- It is difficult and expensive to maintain and repair, representing a significant proportion of total network operating costs.
- While the rest of the network is being modernized, the metallic access network is falling behind in terms of reliability and transmission performance. It is fast becoming the 'weak link'.
- There is no diversity of routeing — a fault on a pair will almost always affect the service to the customer.

- New services (such as ISDN) are being introduced on to the network and these often require a higher level of transmission performance than a simple telephony service.

It is clear that consideration needs to be given to the effective maintenance of the metallic network. The location of network faults is usually the single most significant cost incurred in its maintenance and therefore an area where potential cost savings are large. This chapter looks primarily at the process of locating faults on metallic twisted pairs, first giving an overview of the problem and then discussing some ideas for improving on the methods currently used.

6.2 CURRENT FAULT LOCATION TECHNOLOGY

The presence of a line fault may be discovered either by a fault report from the customer or by a routine test. Measurements of the line may then be made using a remote line test system (LTS) operated from the fault reception office. These measurements are used to estimate the location of the fault in the network so that an appropriately skilled repair engineer may be assigned to its repair. The engineer will then attempt to locate and repair the fault using portable fault location equipment.

Typically the LTS will estimate the fault to be in one of the following locations:

- exchange;
- local network cable;
- customer's premises.

The definition of these three sections of the network is illustrated in Fig. 6.1. The demarcation points are the main distribution frame (MDF) in the exchange and the distribution point (DP).

Figure 6.2 shows a typical LTS architecture. A central processor is connected to remote test heads located in the local exchanges. Operators of the system communicate with the central porcessor which in turn instructs the appropriate test head to perform measurements. These measurements are returned to the central processor where they are analysed to produce the fault location estimate.

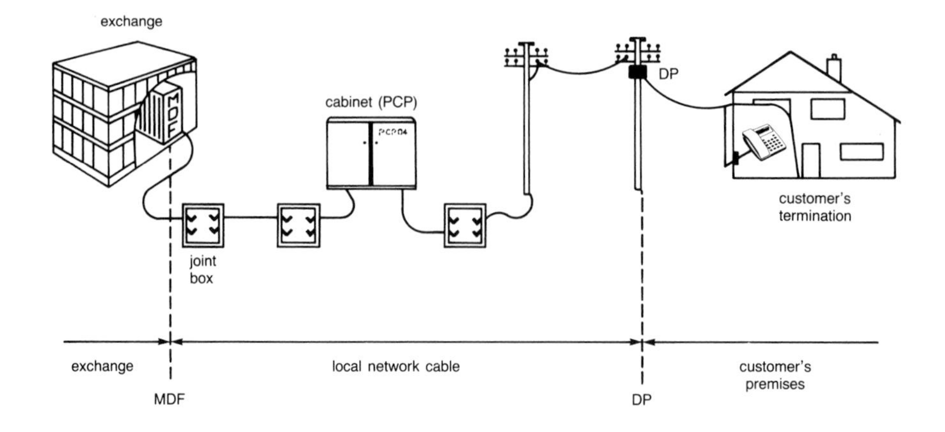

Fig. 6.1 Typical construction of a metallic pair access network connection.

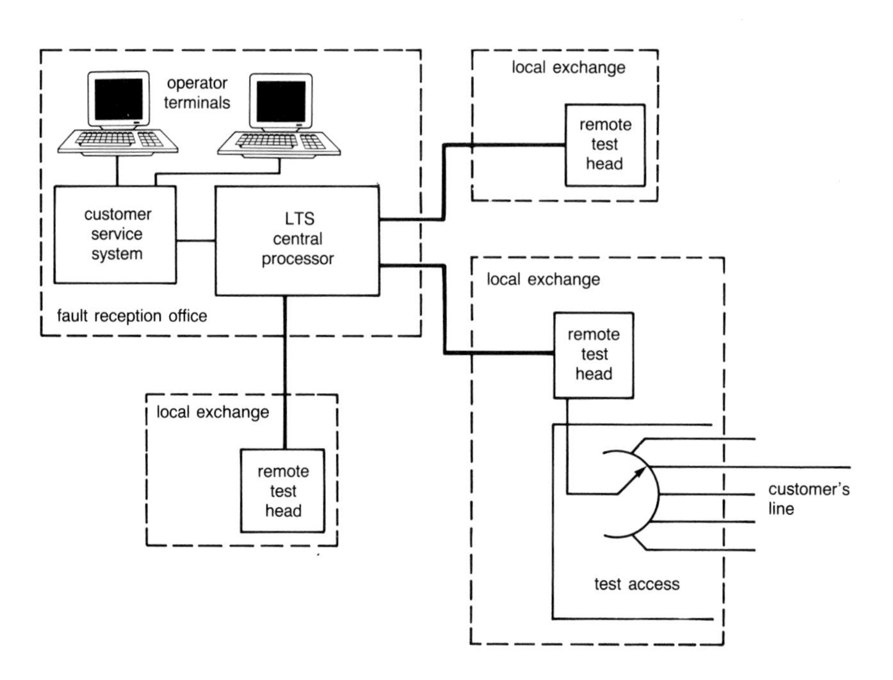

Fig. 6.2 Architecture of a typical remote line test system (LTS).

There are many ways in which current fault location technology is not ideal. Some of the more significant problems are given below.

- The LTS is frequently unable to determine correctly which part of the network contains the fault. Each time it is incorrect an inappropriately skilled engineer may be assigned to the fault's repair and a large cost penalty incurred.
- Existing exchange-based fault location systems do not provide a level of accuracy that enables an engineer repairing a cable fault to proceed directly to the fault without further testing using portable equipment. Portable equipment, because of its limited range, requires access to the cable at several points. This is time-consuming and can create further faults as it involves mechanical interference with the network.
- Existing test systems do not give a geographical location estimate which may be used by the repair engineer. The engineer must use information about the routeing of the cable under test together with the test system output to derive an estimate of the fault's geographical location. This is often a manual process involving paper-based plant diagrams and hence is time-consuming and prone to error.

Portable fault location equipment is a mature area of technology, most of it having been in use for many decades. The LTS however is relatively recent technology and there is still considerable scope for improving it and the way it is used. The remainder of this chapter focuses on ideas for how this improvement might be achieved.

6.3 DESIGNING AN LTS

The overall objective for an LTS is simply stated — it is to obtain information relating to the fault and produce from it, in an acceptably short time, the best possible estimate of the fault's location.

A generalization of the LTS fault location process is given in Fig. 6.3. The first stage is to obtain information specific to the fault. This is the 'fault information' and may come (for example) from the customer's fault report or measurements made of the line. This fault information, together with general information about the network (the 'general network information'), such as line routeing data or details of previous faults, is then passed to an interpretation system which uses it to produce the fault location estimate. The process of designing an LTS therefore has two stages:

- establishing what information can be made available to the system;
- deciding on the best way of interpreting this information.

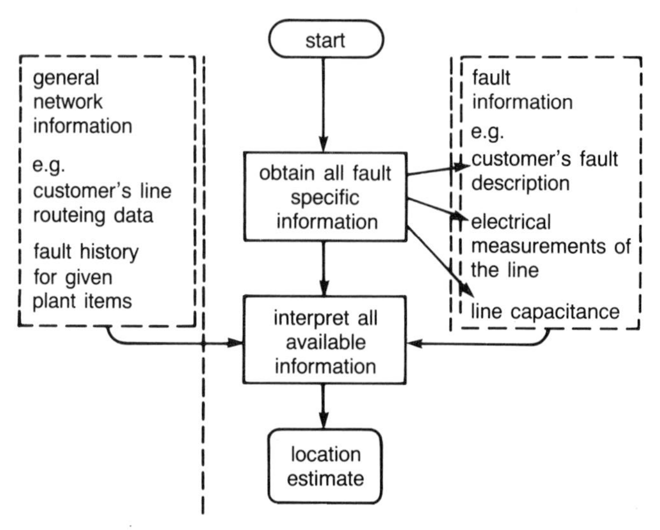

Fig. 6.3 The process of estimating a fault's location.

6.4 WHAT INFORMATION IS AVAILABLE?

As illustrated in Fig. 6.3, the available information is of two distinct types — that which is specific to the fault and that which is general network knowledge. The latter is information about the network that is available at all times whether or not a fault has occurred. This distinction is important as it means the information may be gathered over a long period of time and the cost of obtaining it and making it available to the LTS is spread over all faults.

6.4.1 Examples of fault-specific information

6.4.1.1 Customer fault report

The first example of fault-specific information is the information obtained from the customer. In many cases this will be of considerable use in locating the fault — an example might be a report of frequent incorrect called numbers which would suggest a problem with the telephone's dialling mechanism.

6.4.1.2 Electrical characteristics of the line

Impedance measurements (usually resistance and capacitance) and a.c. and d.c. voltage measurements may be made at the exchange termination. These will give information about the condition of the line — for example, low resistance from one leg to exchange ground would suggest a breakdown of the cable insulation.

6.4.1.3 Time domain reflectometry (TDR)

TDR is currently only used in portable equipment but it has potential for use as a source of information for an LTS. Exploring this potential is a new field of work which is worth considering in some detail.

The principle of TDR is illustrated in Fig. 6.4. An electrical stimulus (V_s) (usually a simple pulse) is applied to the line and the resultant voltage at the point where the stimulus is applied is observed with respect to time. The stimulus will propagate along the line towards the customer's termination. If there is an impedance discontinuity (Z_f) at any point on the line then a reflection of the stimulus will occur at that point which will propagate back towards the exchange. The voltage observed at the exchange is the sum of the applied and reflected signals. The stimuli propagate with a finite velocity which means that the applied stimulus will be separated from the reflected stimulus in time. The time difference or delay (d) between them will be a function of the length of line (x) between the point at which the stimulus is applied and the point at which the impedance discontinuity occurs. If the impedance discontinuity Z_f is caused by a fault the delay between the two stimuli provides information about how far from the exchange the fault is located.

This technique has been understood and used in portable cable fault location systems for many years. There are however problems with adopting the technique for use in LTSs.

- Current systems rely on visual inspection of the response to obtain the delay value. This is not appropriate for LTSs which are intended to be fully automated systems.

- Current portable equipment has limited range and accuracy owing to the visual inspection process. To determine the delay of the fault echo with adequate resolution it is necessary to use high-frequency stimuli (of the order of several MHz) which suffer high attenuation on local network cables designed for audio band transmission. An LTS implementation of TDR would need to use lower-frequency stimuli to ensure it has a range which would cover all customer lines.

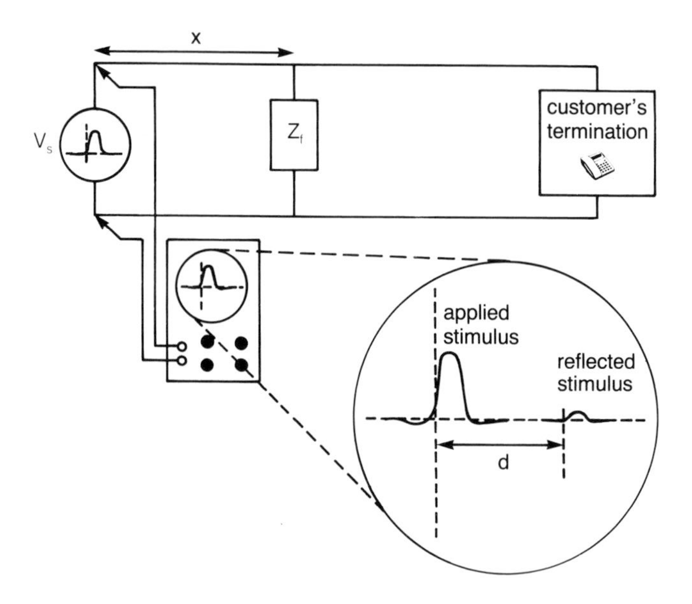

Fig. 6.4 The principle of time domain reflectometry (TDR).

The solutions to these problems may be found by using an analogue-to-digital converter (ADC) to capture the time-varying response digitally and then using digital signal processing (DSP) techniques to process it.

Matched filtering

The problems of reduced resolution and automated identification of the fault echo may be reduced by using matched filtering. This entails looking for features similar to the transmitted line stimulus in the measured response — i.e. correlation between the line stimulus and the response. The resulting filtered response will have a peak at delay 0 due to the transmitted stimulus and further peaks at any delay value where a reflection is present. Figure 6.5 shows the unfiltered measured response and Fig. 6.6 the same response after matched filtering.

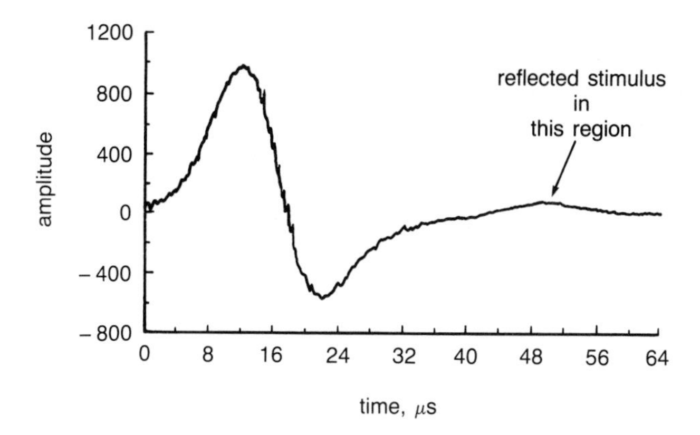

Fig. 6.5 An unprocessed TDR response recorded digitally using an ADC.

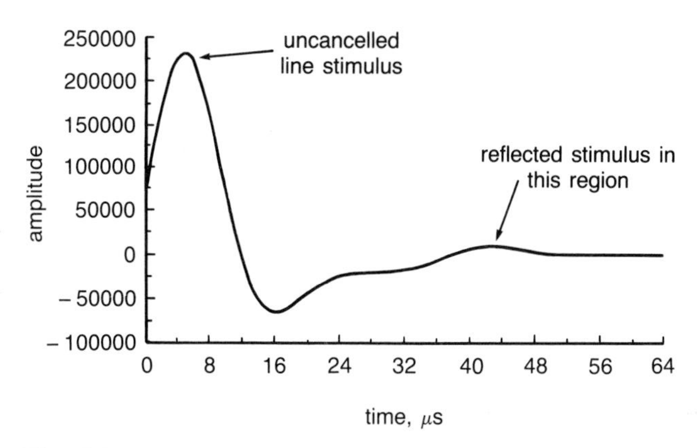

Fig. 6.6 The response shown in Fig. 6.5 after matched filtering.

The reflected stimulus is just visible in the unfiltered response, but determining its delay would be difficult since it is necessary to identify the delay to the start of the reflection (not the peak) and this is very confused by the noise. After matched filtering, not only is the noise level reduced but it is now the peak of the reflection which gives the delay. This lends itself more readily to automation.

Transmit stimulus suppression

This is a method for reducing the distortion of the reflected stimulus caused by the transmitted stimulus by trying to remove the transmitted stimulus from the measured response. In Fig. 6.5, the reflected pulse is distorted by the exponential tail of the transmit pulse. This problem occurs owing to the long duration (i.e. low-frequency) transmit stimulus used. In order to understand how the transmitted stimulus may be removed from the response it is necessary to consider the measurement configuration used when making TDR measurements. This is shown in Fig. 6.7.

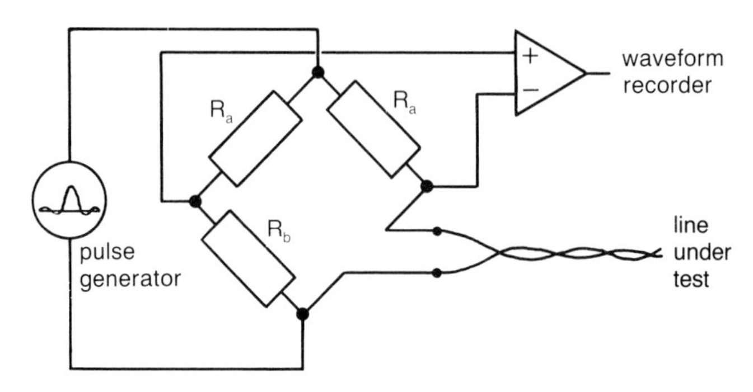

Fig. 6.7 TDR measurement configuration.

The voltage observed by the ADC will be the difference between that across the line and that across the balance resistor, R_b. If R_b is set to a nominal value representing the characteristic impedance of local network lines some cancellation of the transmitted stimulus may be achieved. In practice only a small degree of cancellation is achieved since the characteristic impedance of a local network line is a complex function of frequency which cannot easily be modelled by a simple RLC circuit. The response shown in Fig. 6.5 was measured using this measurement configuration and it can be seen that the transmit stimulus is still by far the dominant feature.

One solution to this problem is to use a real line rather than a simple resistor as the balance impedance. Ideally this would be a working line otherwise identical to the line with the fault — unfortunately access to such a line is often not possible. Again it is DSP which provides the answer. It is possible to use a digital representation of the working line as the balance impedance by performing the transmit stimulus cancellation process in software. The digital representation can either be a stored previous measurement of the faulty line made when it was working or a modelled representation derived from the measurement of the faulty line. A full discussion of these techniques is outside the scope of this chapter (further details may be found in Trigger and Appleby [1]) but a result obtained using the modelled line technique is shown in Fig. 6.8. It can be seen that the amplitude of the reflected stimulus has been greatly enhanced relative to the transmitted stimulus and the effect of the exponential tail is thus reduced.

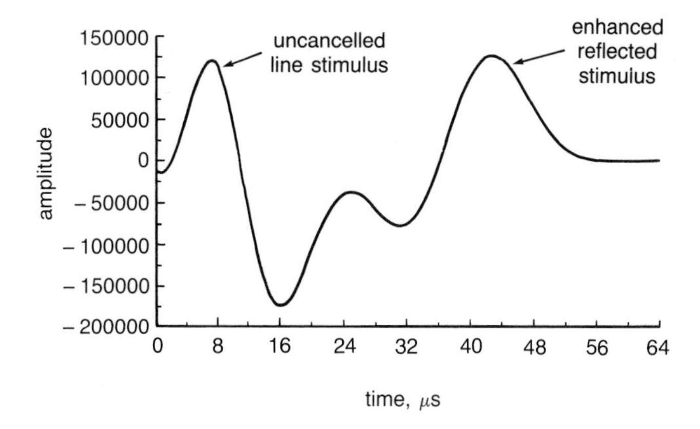

Fig. 6.8 The response shown in Fig. 6.6 after cancellation of the line stimulus using a software model on an identical working line.

6.4.1.4 Noise location

The location of a noise source on a line provides one of the most challenging fault-location problems and any extra information that may be obtained is likely to be of significant value. One possible source of information comes from correlating the noise signal with itself.

Figure 6.9 shows an example where the noise source is between the two legs of the line. In the discussion of TDR techniques it was stated that any signal on the line will propagate along it with a finite velocity and a reflection will occur if the signal encounters an impedance discontinuity. In this case the signal is the noise source itself. This noise signal will propagate both towards the exchange and towards the customer. The customer's termination is an impedance discontinuity and thus the signal present at the exchange end of the line will be the sum of the noise signal arriving directly from the noise source and the signal which has travelled towards the customer's termination and been reflected from it back towards the exchange.

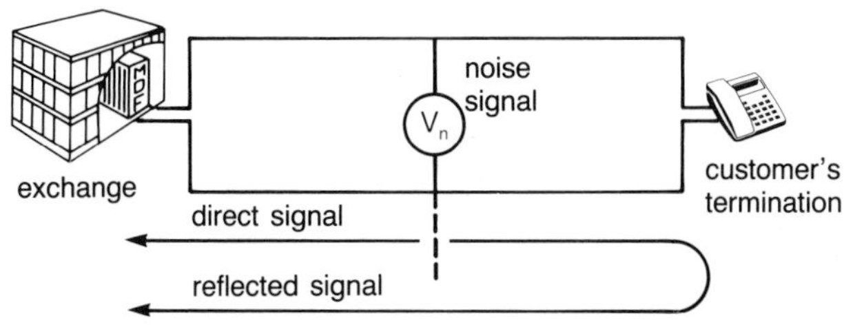

Fig. 6.9 Signals due to a noise source on a customer's line.

Correlating the received signal with itself (using DSP techniques) enables the delay between the direct signal and the reflected signal to be established. This delay is the time taken for the signal to travel from the noise source to the customer's termination and back to the source. It is thus possible to use this delay value to estimate the distance of the noise source from the customer's termination.

6.4.1.5 Other measurements

As well as the electrical characteristics of the line and TDR, many other measurements may be made to provide information about the fault's location. Some examples are a test to check for dial tone availability from the exchange, a test of the performance of the customer's telephone dialling mechanism (made in conjunction with the customer) and the ability of the exchange to provide line-feed current.

6.4.2 Examples of network information

6.4.2.1 Routeing information

A simple example of network information is the routeing data for the customer's line. One of the limitations of current fault-location systems which was mentioned earlier is that they require manual consultation of routeing diagrams before a geographical estimate of the location of a fault may be obtained. Pair-routeing data down to the level of which major plant items (flexibility points and DPs) the line is routed through are often available in computerized local network records and if made available to the LTS could be used to provide the repair engineer with a truly geographical location estimate in terms of the major plant items in the network.

Information about the construction of a customer's line of a much higher resolution than that currently available in electronic databases is usually available on paper maps and plans. If this was also available in an electronic database, the LTS would be able to give its location estimate with much higher resolution than just the major plant items in the network, for example to the nearest cable joint or cable access point. If the graphical map information available on the paper records was also present in the database, then the LTS could provide a geographical location estimate in a 'cross-on-map' form.

6.4.2.2 Statistics of fault occurrence

Over a period of time statistics describing where previous faults have occurred may be gathered and the relative probability of a fault occurring on a particular type of network plant item derived.

6.4.2.3 Fault history

Information about previous faults that have occurred on the line or on an item of plant through which the line is routed could be of use to the LTS. This would allow it to take into account factors such as the likelihood of 'knock-on' faults — faults inadvertently placed on a line during the repair of another fault on an adjacent line. Another example of the use of this information could be in identifying a common cause for several different faulty lines, such as a cable which has been damaged through water ingress.

6.4.2.4 Network expertise

This is network information which currently exists as knowledge possessed by the people who work with the local network, such as those who deal with the faults the LTS is unable to handle and the engineers who repair the faults. The more of this knowledge that can be made available to the LTS the more likely it is to be able to produce an accurate estimate of the fault's location. This also allows an individual's knowledge to be used in the location of all faults rather than just the faults dealt with by that individual.

6.5 INTERPRETING THE INFORMATION

The second stage in the design of an LTS is the design of a mechanism capable of processing both the fault and network information to produce the location estimate. There are doubtless many ways in which this processing mechanism may be realized but only one will be considered here — that is to use a probabilistic rule-based expert system. This system has two main components — a knowledge base and an inference engine. The knowledge base contains rules which describe the relationship between any piece of information used by the LTS and the probability of the fault being at a particular location. The inference engine evaluates each rule in turn to produce the final location estimate.

6.5.1 Why use an expert system?

A clue to why an expert system approach is being considered is present in the previous discussion on what information is available to the LTS. In that discussion use was made of words such as 'suggest', 'might', 'estimate', 'probability' and 'knowledge'. These all indicate the true nature of the problem of local network fault location — it is a process where definitive logical relationships and algorithms to map information on to conclusions will not work effectively since the sources of information used are diverse and the information itself often uncertain and incomplete. A purely logical approach to solving the problem, as might be implemented in a more conventional algorithm, is not able to deal effectively with this uncertain or incomplete data. A probabilistic expert system is, however, ideally suited to this kind of problem since:

- it does not rely on the availability of any one piece of information in order to form sensible conclusions;
- it is able to take into account the certainty associated with different types of information.

6.5.2 How the expert system works

6.5.2.1 Customer line model and probabilities

The first thing the system will do is construct a model of the customer's line using information about the line's routeing. The model identifies the set of possible locations for the fault based on the plant items used in the line's construction. An example line model for the line shown in Fig. 6.1 is illustrated in Fig. 6.10.

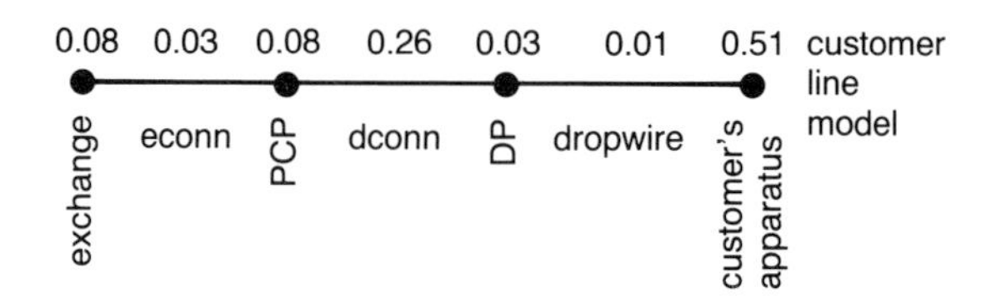

Fig. 6.10 A typical customer line model used by the expert system.

Each location in the line model has a probability value between 0 and 1 associated with it. This probability represents the relative probability that the fault is at that location (it follows that the sum of the probabilities for all the locations in the model will always be 1). The initial values for these probabilities are set by the system using fault rate information derived from statistics of past fault occurrence. Typical initial values are shown in Fig. 6.10.

6.5.2.2 Probabilistic reasoning using rules

The expert system, having set up the customer line model and the initial probability values, processes the information available to it by applying the rules in the rulebase. An example rule will be used to illustrate how the probabilities are adjusted by the application of the rules.

```
IF customer.pcpopened THEN ALTER pcpprob LS 1.1 LN 1.0
```

The rule is in two parts — a conditional part and an action part separated by the word THEN. In the case of the above rule the conditional part is IF customer.pcpopened and the action part is ALTER pcpprob LS X LN Y.

The conditional part of the rule contains some logical expression which will evaluate to one of three possible values — 'true', 'false' or 'unknown'. The logical expression will be a function of one or more of the pieces of information (inputs) available to the system. 'Unknown' will result when one or more of the pieces of information in the expression is unknown. In the example rule 'customer.pcpopened' is simply a boolean indicating whether or not the cross-connection point (PCP) through which the customer's line is routed has recently been opened.

The action part of the rule determines what changes are made to the probability values in the line model. In the example rule only the probability value 'pcpprob' is altered. The values given after the terms 'LS' and 'LN' determine by how much the probability value is to be altered. If the conditional part of the rule evaluates true then the LS value is used to alter the probability; if it evaluates false then the LN value is used. If it evaluates unknown then the action part of the rule is ignored and the probability values remain unchanged.

The calculation of the new probability value from the LS and LN values will involve some rule chaining theorem such as Bayes Theorem. In general, an LS or LN value less than one will decrease the probability, greater than one increase the probability and equal to one leave the probability unchanged. The example rule is thus stating that if the PCP has recently been opened, increase the probability of the fault being at the PCP ($LS > 1$), otherwise leave all probabilities unchanged ($LN = 1$).

After each rule is applied, the probabilities are normalised so that the total probability across all the locations in the line model again sums to unity.

6.5.3 Machine learning

The performance of the expert system described is highly dependent upon the accuracy of the LS and LN values associated with each rule. The traditional method for establishing these values is through consultation with experts in the field of fault location. They would be asked to quantify how much the probability of the fault being at a given location was changed by each piece of information. This process is called knowledge elicitation and is generally accepted as being the most difficult part of developing any kind of expert system. This is primarily because humans are generally good at identifying qualitatively cause and effect relationships (from which are derived the rules in the knowledge base) but poor at quantifying them (setting the LS and LN values in the rules). One solution to this problem is to include a machine-learning system (MLS) in the LTS design which enables the expert system to learn the correct LS and LN values for each rule.

The LTS will produce its estimate of the fault's location which is then passed to the repair engineer as before. The engineer uses portable equipment to find the actual location and then records it so that it is available to the LTS. The MLS component of the LTS will, once a large number of faults have been dealt with in this way, process the actual location information together with its original location estimates to derive new values for LS and LN for each of the rules.

A precise description of the algorithm used to generate the new LS and LN values is outside the scope of this chapter but the essence of it is simple. Consider the example rule given earlier:

```
IF customer.pcpopened THEN ALTER pcpprob LS 1.1 LN 1.0
```

This states that if the PCP was opened recently then increase the probability of the fault being at the PCP (LS is greater than 1) otherwise do not change any probabilities (LN = 1.0). Each time this rule is used in the process of locating a fault then, once the actual location of the fault is known, there are four possible situations which can arise:

- the cabinet was opened and the fault was at the cabinet;
- the cabinet was opened and the fault was not at the cabinet;
- the cabinet was not opened and the fault was at the cabinet;
- the cabinet was not opened and the fault was not at the cabinet.

The respective action that the machine learning system would take in each situation is:

- the LS value would be increased;
- the LS value would be decreased;
- the LN value would be increased;
- the LN value would be decreased.

Using a machine-learning system of this kind, as more faults are processed by the algorithm the LS and LN values should tend towards their correct values and the performance of the system improve. The machine-learning system has the additional advantage of allowing 'suspect' rules (i.e. suspect cause/effect relationships) to be identified by examining the LS and LN values once the MLS has been operating for a long time. If, for example, both LS and LN are very close to unity then it can be deduced that the relationship suggested by the rule is not true.

6.6 AN INTERACTIVE APPROACH

If the initial location estimate produced by the LTS is incorrect the repair engineer is left to locate the fault using portable equipment. The engineer will have access neither to the breadth of information nor to the processing ability of the LTS while performing this location exercise. Figure 6.11 shows how an interactive system might be implemented which continues to assist the engineer until the fault is found. The fault location process as outlined in Fig. 6.3 has been enhanced to include a feedback path via the repair engineer. This might be realized using some remote access device such as a portable terminal and a modem or a radio link. The engineer will inform the system whether or not the fault is found at the location originally given. If not, the interpretation process can be re-run using this new information and hence produce a new estimate of where the fault is. This interactive process can continue until the fault is found.

This approach has an important additional advantage — accurate information about where the fault was found is collected automatically by the LTS since the last estimate it gives will always be the correct one. This will be of significant value both to the MLS and in determining network reliability statistics.

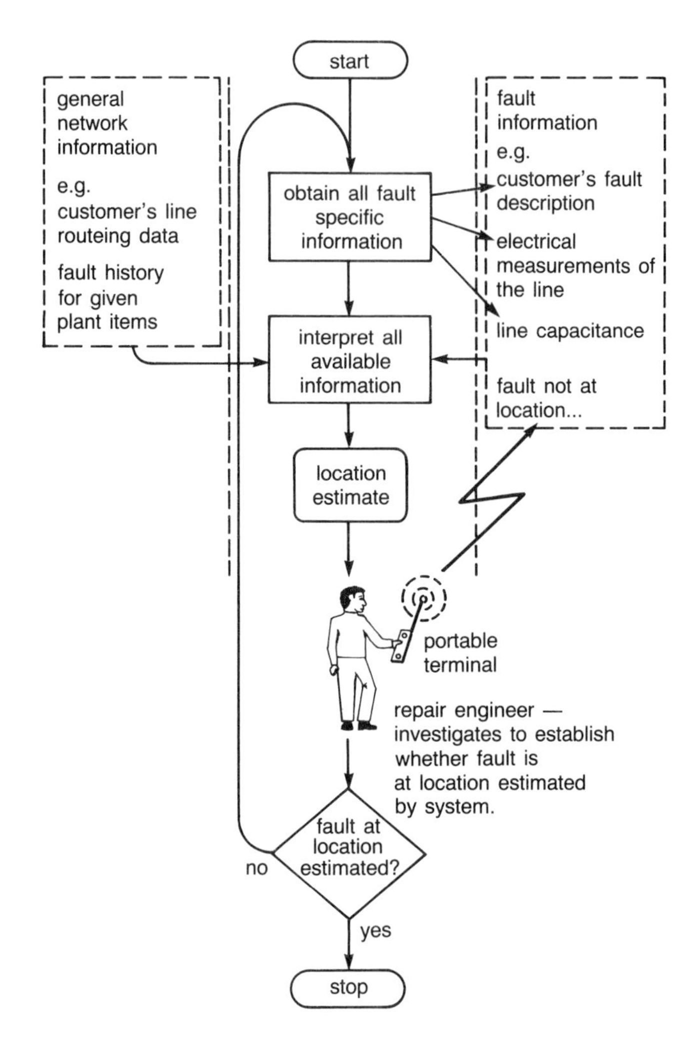

Fig. 6.11 The fault location estimating process enhanced to provide interaction with the repair engineer.

6.7 EVALUATING SYSTEM PERFORMANCE

It is important that LTS development is targeted at those parts of the fault location process where maximum cost savings might be obtained. It is also important that the savings arising from the use of any new technique may be determined and used, in conjunction with knowledge of the cost of deploying the technique, to establish whether there is a sound financial case for making the investment. This section discusses the evaluation of LTS performance.

6.7.1 Choosing a system performance metric

The system performance metric is a measure that can be used to indicate how much better or worse one system is than an alternative system. It must be chosen carefully by considering what it is the system is intended to do. The overall objective of an LTS is to reduce the cost of locating faults — the performance metric that should be used for any LTS is thus the cost of locating all network faults using the system.

There are three functions which need to be determined before the value of this metric may be obtained for any given LTS.

- LTS accuracy — this function is a set of probability distributions giving the probability that, for a fault at any location x on the line, the fault location system estimate will be at location y.
- Cost of LTS error — this gives the cost of locating a fault at location x which has been indicated by the system to be at location y. For example, if a fault in the DP is estimated by the LTS to be in the PCP, what cost penalty is incurred? What penalty is incurred if the DP fault is estimated to be in the exchange? The nature of the function will be determined by factors such as labour costs, work practices and the cost of travelling.
- Fault probability — this is the probability (per unit time) that a fault will occur at any given location x.

6.7.2 Cost metric evaluating using a modelled network

The three functions identified above describe the theoretical approach required to determine the value of the cost metric. It is worth considering how this theory may be applied in practice.

Network statistics may be used to determine the 'fault probability' function and the 'cost of LTS error' function may be determined from labour costs, work practices and so forth. This leaves the 'LTS accuracy' function. A practical method of determining this is to use computer simulation techniques to build a 'modelled' network of faulty lines based on a representative sample of actual network lines. The computer simulation will allow the electrical characteristics, and hence the output from any LTS, to be determined for any faulty line. Any LTS may then be tested on this modelled set of faulty lines to give the 'LTS accuracy' function.

6.8 NON-COPPER SYSTEMS

This chapter has used the metallic twisted-pair local network as its basis for discussing the process of fault location. It is desirable that any system implemented now should be able to accommodate other access technologies as their deployment becomes more widespread — in the interests of effective network management there should only be one system for handling all local network faults.

The expert system approach discussed in this chapter is applicable to any technology, since:

- the customer line model is determined from information contained in line plant databases — any type of plant item, be it based on metallic, radio, optical fibre or any other technology, may be included in the line plant database and hence in the model;
- the system may be enhanced to accommodate a new access technology by identifying the fault-specific and general network information available from that technology and then adding rules to the knowledge base to process that information — the core interpretation process used need not be changed.

Other chapters in this book discussing alternative access technologies indicate what fault-location information might become available from them for the LTS to use.

6.9 CONCLUSIONS

Any telecommunications operator committed to improving the quality of service provided to their customers must concentrate not only on modernising the trunk and junction networks but also on improving the performance of the access network. The rapid and efficient location and repair of faults is of critical importance in achieving this.

This chapter has discussed fault location techniques principally by proposing enhancements which could be made to the current LTS technology. The benefits of investments in these techniques arising from improved fault location capability and hence reduced network maintenance costs and improved access network quality are likely to be substantial. The benefits will increase if the new techniques are designed in such a way that they are able to locate faults on any new technology deployed in the access network.

REFERENCE

1. Trigger J and Appleby S: 'Advanced techniques for customer line fault location', BT Technol J, 7, No 2, pp 30—43 (April 1989).

7

LOCAL NETWORK DISTRIBUTION PRACTICES — A WALK FROM FRAME TO CUSTOMER

D I Monro, D S Butler and S Worger

7.1 INTRODUCTION

As telecommunications engineers look towards a local network in which optical fibres are taken ever closer to the customer it is only too easy to overlook the wealth of engineering ingenuity and expertise that has gone into the development of the copper pair network over many years. It behoves us to look carefully at this achievement so that the engineering efficiency and reliability of the network can be still further advanced, and that the often too sanguine expectations of optical systems engineers can be viewed in a realistic perspective. This chapter attempts to establish such a perspective by viewing the current state of development of the copper network.

The investment in the BT local network is huge. There are over 6900 local telephone exchanges [1] in service which are linked to the customer by the following means:

- by 34 million pairs radiating from exchanges over copper cables varying from 4800 pairs to 2 pairs and some optical fibre cables;
- passing through nearly 250 000 km of underground duct;

- connected via more than 74 000 primary connection points (cabinets);
- jointed in over 3.7 million manholes and joint boxes;
- with final distribution to over 23 million customers via 3 million distribution points and 4 million telephone poles.

The nature of the local network is such that cables fan out radially from the local exchange through a variety of cable types and connection points as shown in Fig. 7.1.

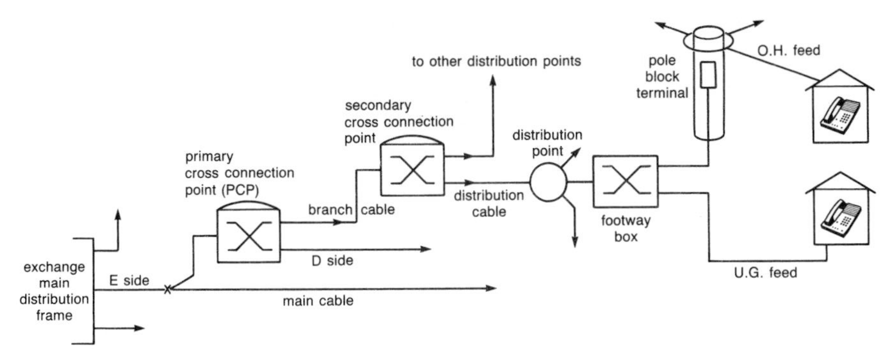

Fig. 7.1 Simplified local distribution network.

Copper cable construction has evolved over many decades and there are many different types of cable which are no longer manufactured but remain in service.

In the late 1960s, aluminium cables were introduced as a cost saving measure because of the rapidly rising price of copper; but today the price advantage has reversed in favour of copper again.

It is now well known that aluminium cables present operational problems due to corrosion, wire breakage and the need for special connector systems.

Copper cables are reliable and offer a simple means of providing 'ringing and speaking' to the customer. Copper cables do not have a high craft dependence for successful installation but nevertheless skill and care are required by the external engineering teams supported by the appropriate connectors, cable closure systems, tooling and mechanical aids.

In BT terminology, cables leaving the exchange are known as 'E side' cables and terminate at a primary connection point or PCP. Cables leaving the PCP are known as 'D side' or distribution cables. The 'E side' cables leaving a large exchange cable chamber can contain up to 4800 pairs. They are protected by an internal aluminium moisture barrier [2] and pressurized with dry air to prevent the ingress of water and the 'D side' cables are filled with petroleum jelly, to minimize water ingress in service.

As the route progresses towards the customer the cable is subdivided and jointing technology is required to achieve this. At points along the local network, street-mounted PCPs (cabinets) and, on some routes, secondary cabinets are used to provide interconnection facilities and flexibility. An 'E side' cable may be split at the PCP and only a proportion of the cable pairs will appear within the cabinet. The remaining pairs will appear in a cabinet further down the cable route.

The diameter of the conductor used for a cable varies. The smallest diameter used is 0.32 mm and cables nearest a large exchange are likely to use this diameter to optimize duct space utilization.

As the cable fans out from the exchange the conductor diameter can be increased up to a maximum of 0.9 mm giving a measure of compensation for the transmission loss to the more remote customers connected to the local exchange. The average route distance [3,4] from a local exchange is 1.7 km representing a radial distance from the exchange of 1.2 km. Less than 10% of all local lines exceed 3 km in length and virtually none exceed 7.5 km in length. At the end of a particular route the final cable distribution is provided by small cables in the range 2-100 pairs. The pairs are connected to the customer either by overhead drop wire from a telephone pole or by an underground feed from a footway box let into the pavement.

Installing a cable requires cable jointing along any particular route. Joints are necessary where cables subdivide and there is a limit to the continuous length of cable that can be pulled into a duct due to friction between duct and cable, the tensile strength of the cable and the continuous length of cable that can be supplied and handled on a cable drum.

A cable joint consists of the individual conductors of one cable electrically joined to the individual conductors of the other cable, e.g. a 4800 pair cable joint requires 9600 individual connections to be made. These have to be electrically insulated from each other and must remain electrically transparent and stable for a service life of at least 20 years in what can be a hostile environment.

For speech band operation the cable run to the customer has to provide adequate transmission and signalling capability. Typically, the worst-case cable attenuation is about 21 dB at 1600 Hz between two telephones working through the same local exchange. The maximum loop resistance is limited to 1800 Ω for loop disconnect or multifrequency signalling.

Operation of copper-pair cables in the local loop is possible at higher frequencies, for example, to provide ISDN (2B + D) and pair gain services. The limiting factors here are cable impedance irregularities and crosstalk performance. For some circuits transverse screened cables are used for 2 Mbit/s operation.

The reality is that the copper network system is a series of cables, joints and connections. This chapter will look at some of the cable connection systems used in the local network and how they are appraised to ensure a reliable and effective copper distribution network and also look at some mechanical and civil engineering aspects.

7.2 JOINTING TECHNOLOGY

In the past, cables contained paper-covered unit twin (PCUT) copper conductors. These were jointed by twisting two bared copper conductors together — the 'crank handle' joint — and the connection insulated with a paper tube. The method was reasonably reliable, although the maintenance of a low-resistance connection was somewhat dependent on the breakdown of the oxide film that formed on the conductors by the application of the system voltage. For trunk and junction work a much higher reliability was required and the conductor joints were twisted and tip soldered. The main problem in jointing cables was the time — weeks in the case of the larger cables — that it took to make a cable joint, resulting in very high jointing costs.

Paper-insulated conductors are now obsolete for telecommunications work and new installations use polyethylene-insulated conductors in polyethylene-sheathed cables.

Whereas the paper insulation was easy to remove from the conductor by hand, the polyethylene insulation needed special stripper tools; consequently the twisted-joint method of jointing was no longer a feasible proposition. Mechanical connector systems were required to increase the speed and reduce the cost of jointing cables.

7.3 MECHANICAL WIRE CONNECTORS

One of the earliest crimp connectors used by BT for cables with less than 100 pairs, and for connections in cabinets, is the 'B wire' connector developed by Bell Laboratories [5] in the 1960s. The BT version is designated the CWI 1A (connector wire insulated 1A). This connector is designed to connect two paper- or polyethylene-insulated copper conductors without the need to strip insulation.

The connector has a thin tin-plated phosphor bronze liner, of oval section, with sharp multiple tangs on the internal surface — rather like an inside-out nutmeg grater. The liner is contained within an oval section brass outer which is itself housed within a plastic sleeve. The connector is filled with petroleum jelly to reduce the risk of corrosion.

The unstripped conductors to be connected are inserted into the connector and the latter is pressed flat on to the conductors by means of a special tool.

The sharp tangs in the liner penetrate the insulation and make several contacts with the conductors. This type of connector is correctly described as an insulation-piercing connector (IPC).

One advantage of the CWI 1A at that time, although it was never designed for such use, was the fact that this connector could be used to connect the aluminium conductors which were then being introduced which could not be connected by a twisted joint because of the natural oxide film on the conductor.

This connector is still in use today, albeit in small quantities, mainly for use on 'difficult' aluminium cables. It suffers from corrosion, despite the jelly filling, and increasing electrical resistance with time.

For small cables, under 100 pairs, the CWI 1A has been superseded by an insulation displacement connector (IDC), designated CWI 8, A, B or C depending on the version. Like the CWI 1A, the CWI 8 connector is for jointing two insulated connectors and is grease-filled. On closing the housing by means of a special tool, the wire is forced into the tag slot and the inside faces of the slot push up and through the insulation and into the surface of the conductor giving two contact points. The slot sides, being cantilevered from the tag base, are sprung apart by the conductor and the elastic deformation maintains the contact force. Like the CWI 1A, the CWI 8 series had to accommodate both aluminium and copper conductors and to achieve this the design of the IDC is critical. Some versions need to be indium-plated to maintain reliability with aluminium conductors.

For jointing larger paper- or polyethylene-insulated cables, over 100 pairs and pressurized, a semi-automatic jointing machine and a crimp connector were developed by BT Laboratories (BTL).

The BT Machine Jointing No 4 [6,7], introduced in the early 1970s and still in service, produces very reliable results and the machine productivity can be spectacular — up to 300-400 pairs/hour. In operation, the jointer lays the conductors to be joined in guides and a hydraulically operated ram closes a connector wire insulated No 6 (CWI6) over the conductors and trims off the surplus wires. Again, the introduction of aluminium cable meant that the CWI 6 connector had to cope with both copper and aluminium [8] and it was one of the few connectors specifically designed by BTL for this purpose. The connectors are contained in a cassette and automatically fed into the crimping head. Any new jointing machine development will have to match or exceed this performance. Figure 7.2 shows the component parts of the CWI 1A, 6 and 8A.

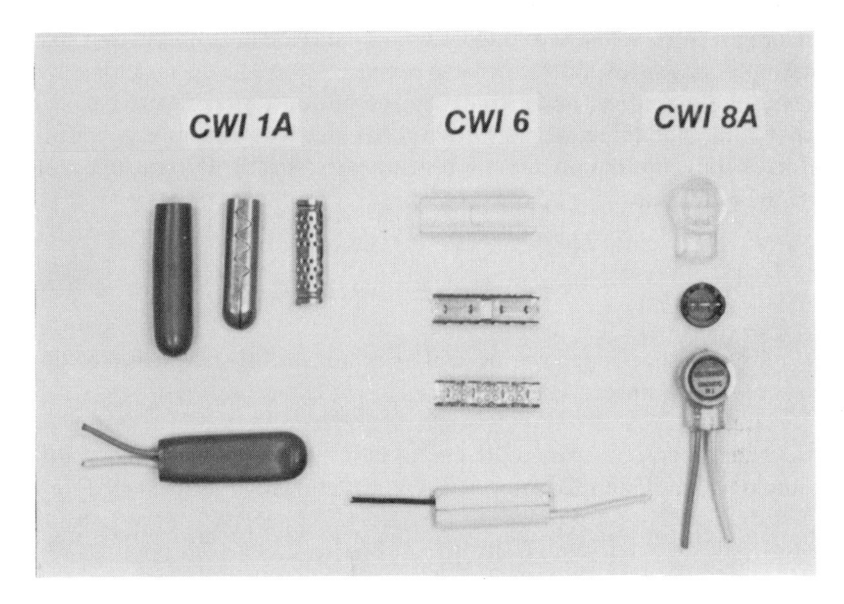

Fig. 7.2 Examples of IPC and IDC connectors.

The most common connector now in service is the insulation displacement connector (IDC). New IDC connection strips are currently being introduced for cabinets, UG jointing and exchange MDFs. The BT 76 series of pole top block terminals use IDCs for both cable tail and dropwire termination. Unlike the CWI 8 which is a discrete connector and can only be used once, many of the IDC strips are required to meet up to 100 re-terminations of varying sizes of conductor and remain reliable. The worst case is the use of an oversize conductor which can deform or splay the IDC tag giving an unreliable connection when subsequently re-terminated with a smaller-diameter conductor. The design of the IDC cantilever is critical for good results and fortunately the shape of a bifurcated tag lends itself to the making of a plastic model and measurement of the forces involved by means of photoelastic methods of stress analysis. Most IDC cross-connect strips cannot be used for aluminium conductors.

Wire wrapping has been employed for exchange connections on distribution frames. This technique uses a square or rectangular post as the connection point and wire is wrapped around the post using a powered wrapping tool. The act of wrapping ‘springs’ both conductor and post and this is locked in by the local deformation of the wire at the corners of the post which maintains the contact force and generally provides a good stable connection. A successful method of wire wrapping aluminium conductors was developed by BTL.

Cold pressure welding techniques used by wire manufacturers to generate continuous lengths of conductor have been explored as a means of in-service jointing. Cold welding, under controlled conditions, gives superb results for copper and aluminium wires without the need to use any connectors. Unfortunately, insulation must be stripped first and its reinstatement is extremely difficult.

7.4 CABLE JOINTING

The jointing of a cable requires several operations which can be broken down into three basic stages:

- cable preparation, where the parent cable conductors are exposed and dressed for the application of the connectors;
- the conductor jointing operation which can be semi-automatic using the machine jointing No 4 or manual using IDC connectors;
- cable reinstatement using a protective sleeve which is coupled to the parent cable.

For lead-sheathed cables a lead sleeve was wiped on to the parent cable forming a high-integrity restoration. With the advent of polyethylene-sheathed cables new methods were required to reinstate the cable sheath. The reinstatement process must be strong mechanically and impervious to water.

Two methods are in use. The first, and preferred method, is injection welding where a chemical bond is made between the sleeve and the cable sheath by injecting molten polyethylene around the interface of a preformed sleeve and the parent cable forming a permanent bond to the cable. This type of closure is extremely reliable but requires a skilled operator for good results. Injection welding has a high labour overhead and the equipment is expensive.

The second, and more popular type of closure, is the heat-shrinkable device. Heat-shrinkable closures are based on irradiated polymers (polyolefins) [9]. Polymers are made up of long hydrocarbon chains randomly distributed within the material. At various points these chains lock together providing most of the material strength. At high temperatures the chains are no longer connected and the material will flow or melt. Manufacturers will decide on the exact composition of the polymer for a specific application. The material is then moulded into the final shape required for the closure and can take the form of simple preformed tubes, wrap-around structures or very complex shapes. The final shape of the closure is then irradiated with high energy radiation in the 1-4 MeV region. This energy causes some of the hydrogen atoms to break free from the chain and where these have been removed polymer chains close together will permanently link. This is known as cross-linking[1] and the mechanical and electrical characteristics of the material are permanently changed so that it will not flow or melt at any temperature.

The preform is then heated and behaves like a rubber which can be stretched by a defined amount and is allowed to cool in this state. Large closures are reinforced using aluminium inserts and some reinforcing material can be introduced into the polymer for added strength. This is important for pressurized cables working at 620 mbar gauge (9 psi).

The cross-linking gives the material an elastic memory and the application of heat, in the form of a flame, to the closure surface causes the closure to shrink down to the design size on to the cable. Shrink-down ratios of 4:1 or 5:1 are possible when the surface temperature of the closure exceeds about 120 °C.

It is not possible to form a chemical bond to the cable sheath directly and a hot-melt adhesive is used to glue the closure extremes to the parent cable.

A new closure having a non-cross-linked polyethylene lining has been developed in Japan that amalgamates with the parent cable. The result of peel tests and sectioning of the amalgamation line indicates the formation of a very strong bond with the parent cable.

After the parent cable ends are prepared and the jointed conductors have been protected by a paper wrapping, aluminium foil and/or metal braid, the closure is shrunk down using a gentle flame from a propane blowtorch moved constantly around the area of the work. By working from the middle to the end the air within the closure is displaced to the ends avoiding air bubbles and the closure gradually conforms to the required shape.

It is important not to overheat the closure as this can cause damage to the closure and the joint. Some manufacturers use a speckled heat-sensitive paint on the outer surface which disappears when the required surface temperature is reached.

The mechanical strength is preserved by a combination of the hoop stress and the glue line where the closure meets the parent cable and forms a generally reliable closure. Figure 7.3 shows a simulation of the various stages of constructing a heat-shrinkable cable joint.

---

[1] Cross-linking can also be achieved chemically.

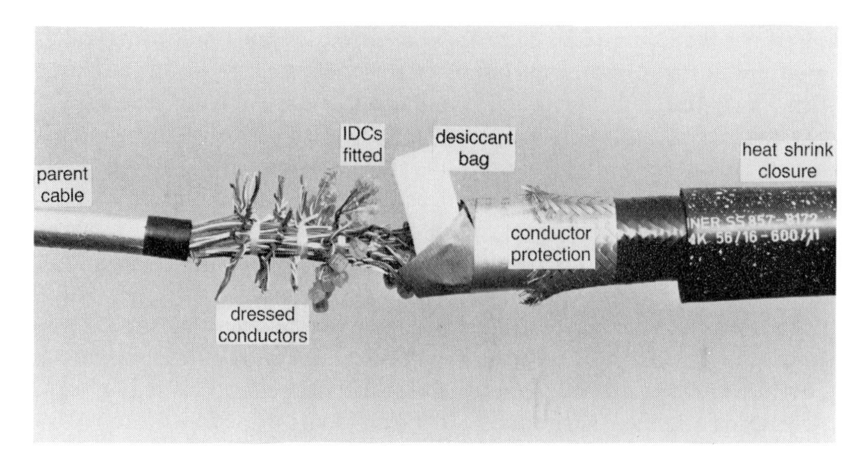

Fig. 7.3 Simulated stages of constructing a cable joint.

The use of a naked flame to shrink down a closure can present unacceptable safety hazards due to the presence of explosive gas in manholes and great care is required in testing for gas, using approved gas detectors, before any closure operation using naked flames is undertaken. Some utilities have developed electrically heated closures to avoid this problem. For telecommunications work there are developments in hand to use electrically shrunk-down closure techniques. If successful and economic, the integrity of an electrically shrunk-down closure will improve compared to the use of a flame particularly in difficult situations such as where the cable is at the bottom or adjacent to the wall of a manhole.

There is a wide variety of closures used within the network and these get smaller as the cable route approaches the customer. To accommodate the cable fanning out factor multi-entry closures are used.

As the cable route moves further down the D side a simple two-part sleeve is employed for smaller diameter cables known, within BT, as the 30 series sleeves. These come in various sizes. The incoming/outgoing cables are secured using a semi-flexible resin poured into the base of the sleeve giving a mechanical termination for the cable and also providing a hermetic seal. Later versions of the sleeve use a heat-shrink band to form the seal between the cable entry points and the cables, thus eliminating the resin seal.

The cables are jointed using IDCs above the seal and a desiccant bag is placed in the top half of the sleeve to provide additional protection for the connectors. The sleeve is then closed using a plastic dome and clamp. This type of sleeve is normally installed in a footway box.

The final part of the distribution network to the customer can take two forms — an underground feed or an overhead feed mounted on a telephone pole. In the latter case up to 20 pairs will be extended from a footway box and terminated on a pole block terminal. Drop cables are mechanically terminated by means of a preformed wire to a ring at the top of the pole used to provide the customer with service. The 'drop cable' is fixed at some convenient point on the exterior of the customer's premises and is electrically terminated at the customer's service outlet.

The pole top block terminal is the component in the distribution network that is exposed to the worst operating environment and early designs had a high fault liability due to the ingress of moisture.

A new design of pole block terminal, the BT76 series, has been recently introduced and uses IDCs to terminate the cable tail and the drop wire. The connections to the tail are sealed with flexible epoxy resin and the drop wire IDC is grease-filled. This new design has overcome many of the earlier problems, particularly corrosion and loss of insulation resistance associated with the older block terminals using exposed screw terminations.

In summary then, it can be seen that the metallic route from the local exchange to the customer is made up of many elements involving different practices and technologies. Although the individual components are relatively simple their performance is crucial to the network as a whole. The quantities of components involved are staggering and some £300m was spent annually on stores alone for renewal and provision in the local loop in 1989. It is vital that these components are fit for the purpose. There are a wide range of connector components available from manufacturers worldwide and many of these were primarily designed for a home market where larger diameter cables are used than in the UK which have a range from 0.32 mm to 0.63 mm diameter. To see if these connector systems are fit for BT use a rigorous engineering assessment and testing regime is used before adopting them for service in BT.

7.5 ENGINEERING APPRAISAL

The main criterion for connectors concerns their mechanical and electrical performance under defined environmental conditions. The transmission line parameters L, C, R and G define the impedance and transmission loss of the cable. The values of L and C per unit length are fixed for a particular cable design. Because the normal operating frequency of a local line is low the influence of a joint on L and C is minimal and the main concern is that a connector does not significantly modify the R and G terms causing an increase in attenuation.

At low frequencies the line attenuation, α, is dominantly controlled by the R and G terms and is proportional to $\sqrt{(RG)}$ to a first approximation.

This can be expressed as $\alpha = k.\sqrt{(Rline/Rins)}$ per unit length where k is some constant, *Rins* is the effective insulation resistance of the line and *Rline* is the loop resistance. Typically connector specifications will call for a change in ohmic resistance of the connector to be less than 2.5 mΩ under all test conditions and a minimum insulation resistance of 10 MΩ. Any variation of the connector parameters R and *Rins* will have some effect on the transmission path.

Tests have shown that the insulation resistance is subject to the most variation and under wet conditions the insulation of block terminals can break down cyclically as conducting moisture paths are established between the A and B legs and then broken down rapidly by the system voltage. In early designs of block terminals the insulation could break down permanently owing to tracking in the insulator. This effect causes 'frying noises' on the customer's line.

BT uses millions of connectors each year. If a connector system with latent faults were adopted for the network the consequence of an in-service failure, or, worse, some form of intermittency giving the 'fault not found' condition can result in remedial work costing a thousand times more than the cost of the failed component. Perversely, the introduction of the digital network has highlighted some of the shortcomings of the local loop in terms of noise due to poor interconnections within the access network.

To minimize this type of problem a testing regime is used based on the British Standards Institution specification BS2011: 'Basic Environmental Testing Procedures' [10]. This is an omnibus specification and relevant clauses are used from this specification together with some special type tests developed by BT to determine the fitness for service of connector systems. These tests accelerate inherent failure mechanisms in components and provide a good pointer to the service life of the component. The list below is an abstract of BS2011 Pt2.1 and gives examples of the tests used and their application. The references are to clauses in the specification:

- Salt mist (Ka:1982) — corrosion and insulation susceptibility test.
- Damp heat — steady/cyclic (Ca:1977;Z/AD:1989) — composite temperature and humidity cycling test for corrosion, film formation and insulation degradation.
- Dry heat/non-heat dissipating specimens (Ba:1977) — terminals operating at rated current-variation of contact resistance.
- Bump test (Eb:1987) and vibration (Fc: 1983) — fatigue, resonance and shock tests.
- Industrial atmosphere (Kc:1977) — gas burning test for corrosion tests.

Other tests:

- BS3900 Pt F2:1973: Determination of resistance to humidity — cyclic condensation.
- BT Specification D2920: Modified thermal soak/cycle tests for IDCs and IPC connectors.
- Tensile tests — pull out of conductors from connector.
- Flexure tests — to determine effects of bending over defined radii.
- Low temperature — re-termination tests at -10 °C to determine ability of conductor and connector to work reliably at low temperatures.
- Electrical tests — working voltages/currents applied during most tests and IR and CR monitored continuously, flashover tests at 1050 V d.c. for dielectric strength.
- 'Duck Pond' test — terminated IDC connectors immersed in water with copper anode at base of tank, used for determining efficacy of grease filled connectors.

All of these tests are devised to provide data on the expected reliability of connector systems and to enable defective products to be identified as early as possible in the testing programme.

The salt mist and gas burning tests are not entirely satisfactory in that the results tend to be severe and inconsistent; however, in some tests they can give very good correlation with service conditions. In one example the corrosion after five years exposure was indistinguishable from the result of a 21 day salt mist test on a product that was causing operational problems. The salt mist test was originally devised for testing the stability of paint films and later adopted for telecommunications components. In general, the salt mist test tends to induce catastrophic failures in connectors under test and a more representative and gentle test is required where a component can degrade gracefully.

One test which has been developed recently by BTL, based on BS3900 Pt F2 [11], is the saturated atmosphere test. In this test the component is placed in a 100% water-saturated atmosphere cycling between 42-48 °C every hour for 21 days. As the temperature is cycled, water will condense on the component under test at the lower temperature simulating real-life conditions. The insulation resistance of individual connectors is measured continuously during the test with 50 V d.c. applied to the A and B legs. Initial results show that the insulation fails gently over days, rather than catastrophically in hours as in the salt mist test, and valid comparisons can now be made of different connector types. It may replace the salt mist test for future BT work.

The gas burning test (BS2011 KC:1977), which is used for corrosion susceptibility, was originally developed by BT, but has been frequently criticized for lack of consistent results. An alternative test has been proposed and developed by Prof W Abbott of the Battelle Institute of Columbus, Ohio, USA; it shows considerable promise in simulating real in-service corrosion effects over short test periods. This test, known as a 'flowing mixed gas test' (FMG), is currently being studied under a UK Department of Trade and Industry initiative by several laboratories in the UK including BT.

The essence of the test is that minute amounts of reactive gases, having concentrations measured in parts per billion (ppb), are mixed with air and introduced into a chamber under controlled conditions of temperature and humidity. The gases used are chlorine, hydrogen sulphide, and nitrogen dioxide. The reaction kinetics are complex and outside the scope of this paper. More information on this method of corrosion testing is available elsewhere [12]. The FMG test is showing good correlation with real field data obtained by many years of component exposure in different hostile environments. Inter-laboratory comparisons indicate that this test has reproducibility and can be analytically controlled. The concentration of gases used varies according to the severity of test required. Table 7.1 below gives the composition of the FMG for various degrees of severity.

Table 7.1 Flowing mixed gas.

Gas concentration — parts per billion				
Severity	H_2S	Cl_2	NO_2	*RH% T°C*
1	—	—	—	—
2	10	10	200	70/30
3	100	20	200	70/30
4	200	50	200	70/50

Notes: Severity 3 is generally used for BT work.
Gas flow rate 2-4 l/min, RH controlled to ±2%

The analytic equipment and control systems required to maintain these very low gas mixture concentrations are complex. Experience has shown that with correct maintenance routines and calibration techniques the FMG test is a viable one for a testing laboratory.

In practice, small metal coupons (e.g. gold, silver, nickel and copper) are introduced into the chamber along with the sample under test. The coupons can be measured for weight gain and also, by using an electrochemical reduction method (cathodic reduction) and other surface analysis techniques, the oxides, sulphides and chlorides formed on the token can be quantified. In this way excellent quality controls can be placed on the test. Coupons can also be cut and one half sent to another collaborating laboratory for independent analysis as part of the quality control audit. The Battelle cabinet represents a significant step forward for testing electrical components. More work is required before this test is adopted as a national standard.

Thermal testing can show up potential problems due to thermal expansion and contraction and stress relaxation in connector systems. Vibration and bump tests can often show up weaknesses in connector design or application. Structure-borne vibration can be transmitted to a connector system causing mechanical resonances to occur which can fracture the conductor where it enters the IDC or dislodge it from the IDC. To minimize this effect the conductors are taped down during the cable joint construction damping out the vibration.

This brief description of connector testing shows some of the techniques used to appraise a connector system. Many other tests are applied including making up specimen cable joints. Much of this work is repetitive, tedious and time-consuming but is essential to identify the right connector components for use in the network.

7.6 DATA ACQUISITION

A UNIX-based measurement system is used to capture electrical data for connector change of resistance and insulation resistance between adjacent *A* and *B* legs of block terminals. The primary data is presented in graphical and tabular form. The connector change of resistance test uses a four-wire Kelvin method with a resolution of 10 $\mu\Omega$. To eliminate the resistance variation in the connecting tails after environmental testing the samples are returned to a temperature-controlled room before measurements are made.

Connector insulation resistance (IR) can be measured continuously during environmental tests. For connection strips and block terminals each *A* and *B* leg is polarised by a current-limited 50 V supply. Higher potentials are sometimes used for special tests. A precision resistor is connected in series with each terminal pair and the voltage across this resistor is measured at intervals during the test. The insulation resistance is derived from this voltage.

Figure 7.4 shows the result of insulation testing for a set of six block terminals tested at the same time in the test chamber using the BS3900 cyclic condensation test. Three of the samples maintained an extremely high insulation resistance during this test and are satisfactory.

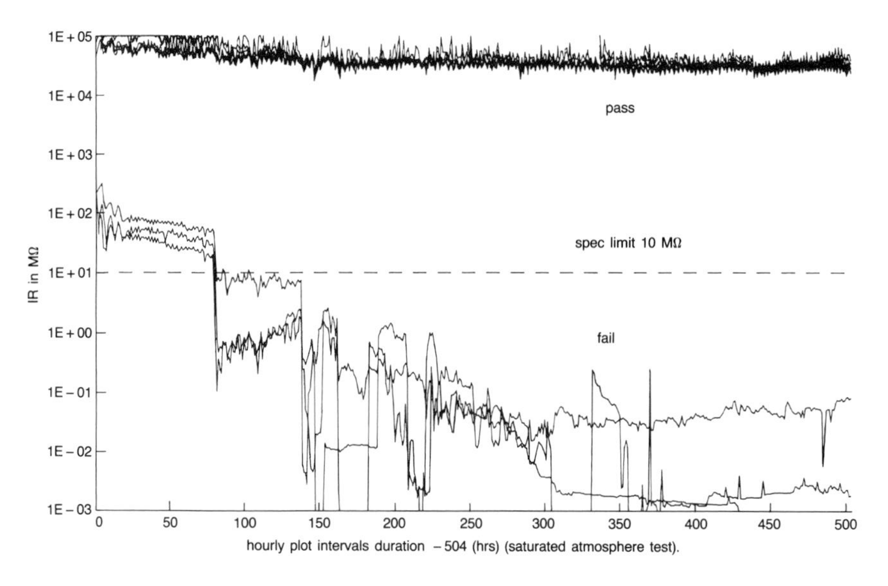

Fig. 7.4 Variation of IR of block terminal during environmental test.

Three of the samples failed to pass the 10 MΩ IR limit. In one sample the IR flattens out to a value of 100 kΩ and the remaining two show violent fluctuations of IR. One sample eventually becomes a virtual short circuit by the end of the test due to carbonisation of the insulator. These block terminals would be service-affecting if installed in a local network.

7.7 SHEATH CLOSURE TESTING

Closures developed for the local network have to withstand severe environmental conditions. The climatic variations of temperature and rainfall are large. Extremes of temperature in recent years have been in the range −30 °C to +70 °C and above for closures exposed to sunlight owing to solar gain. Closures also have to operate under water because deep manholes can (and do!) fill with water and the closure performance under hydrostatic pressure is important. The operation of a closure in depths of water up to 10 m can be determined in a pressure tank developed by BT which can accommodate up to six closures and subject them simultaneously to hydrostatic pressure and tensile load.

Thermal cycling and accelerated ageing tests have been devised that will simulate the range of in-service conditions. In addition, closures will be subjected to mechanical stresses due to structure-borne vibration from traffic, unstable geological conditions and traffic-induced creep. Torsion, impact tests and static loading tests are used to determine the closures' ability to withstand the working conditions. These may seem very harsh tests to apply to a closure but they reflect the real world and give confidence that a closure which is correctly installed will have a service life of at least 20 years.

One severe test is temperature cycling in water. In this test closures are pressurized and placed in a 1500 litre water bath and alternately subjected to water maintained at 4 °C and 60 °C by means of two temperature-controlled 3000 litre capacity water transfer tanks. Where required the closure is pressurized to 620 mbar and the internal pressure of each closure under test is monitored using a small silicon pressure transducer connected to a data logger. Any loss of internal pressure during this test represents a failure.

For non-pressurized closures the internal humidity, corrected for temperature, is measured over a period of time. An increase of 10% rh within the closure is regarded as a failure. It is usually found that the failure is catastrophic.

In the course of an investigation unexpected effects can also occur and these are discussed with the manufacturer. Field reports are also very important because no system of testing can ever fully simulate real world conditions. Testing is however very important as a first line of defence in preventing unsuitable components being introduced into the network. The result of this work is to generate specifications which detail the performance expected from a connector or closure system to meet the network requirements at an economic price.

New demands are being placed on manufacturers and, with a changing network, new connection devices and closures are required. In this context the emerging optical network will require a new generation of closures to protect the fibres.

The ability to negotiate with a supplier depends on being an informed customer. One consequence of testing is that the customer knows what products not to buy and it often enables an existing product to be improved by working with the supplier. Buying the right product for the external network means a reliable and sustainable local distribution network can be achieved.

7.8 CIVIL ENGINEERING ASPECTS

In cities and towns all of the utilities bury their plant underground (UG) wherever possible. The various services are often installed under pavements or in the highway. Over the years from the time gas mains were first laid in the 1800s the layout and position of utility plant has become something of a jungle. As local authorities and utilities have changed over the years records have become lost or were never recorded! Utility plant is generally buried at different depths with water being the deepest and electricity and telecommunications cables nearest the surface. Services are now being colour coded to aid identification. Utilities, including BT, provide plans of their plant to aid contractors. The excavation of a pavement or highway is potentially dangerous and expensive. Utility records are being upgraded using digital mapping but this will take time and much plant is still unrecorded [13]. BT has a major task in hand of translating its cable plans to digital mapping. In practice the current accuracy of plant maps is only fair and cable-location devices based on low-frequency transmission and the detection of 50 Hz fields are used to locate UG plant. These suffer from field distortion due to adjacent plant degrading the spatial and depth resolution and must be used with care. Electric power cables present a major hazard and the radiated 50 Hz magnetic field is weak.

The National Joint Utilities Group (NJUG) publish a specification for the appraisal of cable locators. This specification calls for a locator to detect the magnetic field at a distance of 1 m from a cable carrying 20 mA r.m.s. at 50 Hz — this is equivalent to a flux of 4.10^{-8} Wb/m^{-2}. It is important that the performance of any cable locator is sustained for safety reasons.

Mention should be made of ground probing radars (GPRs) for the delineation of UG plant. Pulsed GPRs work on the same principle as standard navigational radars and are commercially available for geophysical survey work. They can identify metallic and non-metallic utility plant and other artefacts by reflection due to the change in local dielectric constant. The centre frequency, pulse duration and repetition rate used depend upon the application. For the detection of utility plant a high (0.5-1 GHz) centre frequency is used with nanosecond-duration pulses. Reflections at the air/ground interface and the effect of non-homogeneous soils cause difficulties and the derived images from GPRs require expert interpretation in all but the simplest of cases. Much development is required before GPRs can be used by field engineering staff.

Since the earliest days of telegraphic communications wooden poles have been used as a means of supporting the transmission media and poles are still used in vast numbers by many authorities throughout the world today for telephony. In the UK, pole routes are used for local distribution particularly in rural areas and as DPs in suburban areas. The most commonly used pole is from the redwood pine (pinus sylvestris) although other species are sometimes used. For some applications hollow metal or GRP poles are used for safety reasons. The telephone pole population in the UK is about 4.5 million and some 80 000 new poles are purchased annually for new installations and renewal work. The service life of a pole varies considerably and is a function of local biological conditions but typically the service life is about 44 years. Being organic in nature poles decay owing to fungal and insect attack. The most common form of preservative used for protecting the poles is coal tar creosote oil. Various processes can be used to enable the preservative to penetrate the pole, the most common being the vacuum/pressure process where the pole is placed in a pressure tank containing the preservative and treated using a defined pressure/time/temperature cycle [14]. The take-up of preservative will be in the region of 120 kg/m^3. This treatment does not mean that the pole will be immune from fungal attack or other damage.

Within BT a formal pole inspection is carried out every six years; it is the line technician's duty to inspect and test any pole before climbing it at any time.

Many pole-testing methods have been proposed and tried over the years based on ultrasonics, radiography, resonance methods and bending moments but have not been adopted for safety reasons, practicality or cost. The most common test is the hammer test which is simple and analysis has shown that the testing technician always errs on the side of caution.

Drop wires are used to connect the customer from the final distribution point. Normally, the maximum span is 70 m and the effects of wind, icing, mechanical strength, pole loading have to be taken into account when designing this part of the distribution system. The tension, T, in newtons, of a symmetrically suspended overhead wire is given by the expression:

$$T = wL^2/8s$$

where w is the weight, from all causes, of the cable in newtons/metre,
L is the span length in metres, and
s is the sag or dip.

The effective weight, w_e, is $\sim\sqrt{[(\text{weight of conductor} + \text{ice})^2 + (\text{wind load})^2]}$ and the sag will be: $s = w_e L^2/8T$

The wind loading will depend on the aerodynamic properties of the drop wire which will vary with the cross-section profile. Wind tunnel tests have been made of representative sections of actual and possible drop wire designs to optimize the cable section. Ice loading is taken as 5 mm radius for design, but in severe winters ice can build up to over 50 mm. Another issue affecting the design of drop cables is the dynamic breaking strain. One way of determining the dynamic breaking strain of drop wires is to suspend a wire at low level between two poles and drive an adapted vehicle at the wire under test. This simulates real life situations and it is important that the wire fails and acts as a mechanical fuse to protect the pole. Strain gauges and other instrumentation are used to obtain the dynamic breaking signature of the cable. A safe method of determining the dynamic breaking strain — but less fun — is to use a drum driven by a hydraulic motor which winds the cable under test on the drum. The tension in the cable is measured using a load cell attached at the end of the test frame. Very consistent results have been obtained using this method of dynamic testing and may form part of a future specification.

Gas is always a danger associated with underground plant and strict rules are applied for gas-testing procedures before entering any UG structure. Potentially explosive situations can arise from the build-up of methane or propane. Gas detectors are employed based on a Wheatstone bridge with one or more elements which become unbalanced in the presence of gas; the output from the bridge is indicated on a meter which shows the lower explosive limit of 4% and a danger zone. Carbon monoxide and oxygen deficiency in UG plant also present a danger to workmen and a Davy safety lamp is used for deep structures. New forms of gas detectors are under development based on molecular absorption and could herald a new era in portable continuous gas monitoring.

7.9 THE CHANGING LOCAL NETWORK

Copper has reigned supreme in the local network since the inception of public telephone service. New technologies are emerging that will change this situation. The most significant will be the introduction of optical based systems working directly to the customer or acting as a broadband primary highway feed from the exchange and then over copper for the last stage of distribution. The customer will have a choice of narrow or broadband service. The use of radio-based systems forming the last part of the local network distribution is also possible and is particularly useful for providing service in remote areas.

Blown fibre technology [15] allows a broadband fibre connection to be made from a distribution point directly to the customer and has many applications for in-building wiring.

In a competitive environment there is a need for the network operator to isolate his plant from the customer's apparatus for the purpose of line condition monitoring and testing. Remote isolation devices are being studied and could well be integrated with digital mapping to provide a precise location of incipient or actual fault condition.

More intelligence can be built into the local network enabling re-configuration and service provision to be made remotely. Fibre with its intrinsic low-loss transmission would allow a reduction in the number of telephone exchanges and provide a wide range of services. The desire to move into new services may be constrained by the government of the day [16,17].

The technical and commercial viability of these systems has yet to be tested. Implicit in mixed transmission media systems will be the need for active electronic and optical subsystems within the distribution network. The environmental conditions to which these subsystems will be exposed are the same as for copper interconnect components and the assessment methods described are equally applicable for these new components.

It is highly likely that copper will continue to dominate the local network, particularly for domestic customers, into the next century in spite of these new, and technically exciting, developments.

7.10 CONCLUSIONS

This chapter has taken a brief walk from the exchange mainframe to the customer and looked at some of the processes involved in the design and assessment of the component parts of the network.

Although the performance of the UK local network is sometimes criticized, it should be appreciated that modern local network techniques can provide a high degree of reliability under diverse operating conditions, and if the best of new practices can be introduced during progressive refurbishments a very high performance, low whole-life-cost network can be achieved.

The local network is on the threshold of dramatic change and will eventually incorporate many of the optical transmission and switching technologies discussed in this book. It is critical however that the new systems give efficient and reliable performance in the demanding environmental and work-force situation of the local loop. The hard road trodden by the copper network provides salutory instruction on the problems to be faced.

REFERENCES

1. '1001 Facts. An ABC of BT', (1988).

2. Harrison J C: 'The metal foil polythene cable sheath and its use in the Post Office', IPOEE Paper No 229 (1968).

3. Swain E C: 'Local lines — past, present and future', IPOEE Paper 226 (1965).

4. Dufour I G: 'Local lines — the way ahead', BT Eng J, 4, Pt 1, pp 47-51 (1985).

5. Graff H J et al: 'Development of a solderless connector for splicing multipair cable', BSTJ, XLII, No 1 (January 1963).

6. Harding J P: 'The design of a wire jointing machine for subscribers telephone cables', BT Research Dept Report 308 (1972).

7. Harding J P: 'Design of a connector for copper cables', BT Research Dept Report 309 (1972).

8. Hines H E: 'Evaluation of displacement technology for use in BT', BT Research Department Report 794, also BP 1382811 and 1435011 (1982).

9. MacAllister D (Editor): 'Electric cables handbook', BICC/Granada (1982).

10. BS2011, Generic specification, in many parts, for basic environmental testing procedures.

11. BS3900 Determination of resistance to humidity (cyclic condensation) (1973) confirmed (1984).

12. 'Corrosion of electrical contacts — progress in environmental gas testing', NPL Conference papers (November 1988).

13. 'Roads and the utilities', (The Horne report) Dept of Transport HMSO (1985).

14. BS5589: Preservation of timber (1989).

15. Cassidy S A and Reeve M H: 'A radically new approach to the installation of optic fibre using the viscous flow of air', Proc 32nd IWCS, pp 250-254 (November 1983).

16. DTI Communications Steering Group Report: 'The infrastructure for tomorrow' (The McDonald Report) HMSO (1988).

17. 'Broadcasting, the way ahead', HMSO (1988).

Part Three

Optical Systems

8

STAR-STRUCTURED OPTICAL LOCAL NETWORKS

J R Fox and E J Boswell

8.1 INTRODUCTION

The star structure, in which dedicated links radiate out from the exchange to each customer, has been the basis of telecommunications networks since their inception. It is a natural arrangement where a dedicated both-way connection to each party is required. Other communications systems however have adopted different architectures because of the different nature of their services. Thus cable TV, aimed at broadcasting the same package of services to each customer, found the tree-and-branch structure natural; computer and terminal interactions over a LAN, with their sporadic nature of data communications, found shared buses and rings an efficient arrangement.

In fact each type of network structure is capable of adapting to different applications, though with varying ease. With the twin forces of technology push (optical systems, VLSI circuitry, etc) and market pull from the demand for new services, there has been a worldwide study of new broadband multi-service networks and, in particular, a consideration of what structure can best accommodate present and future service demands. Star-structured networks have been a natural focus of attention because of their inherent compatibility with real-time two-way service and their operational ease. However it is necessary to make a balanced assessment of advantages and disadvantages, and these will be examined later in the chapter.

First the possible star architectures will be reviewed, looking at their general application, their features, and their development. The key network features will then be examined with particular regard to the impact of new technology. The broadband integrated distributed star (BIDS) network will be described in depth as an example of how a star-structured network can offer a wide-ranging service package. Finally future development and applications will be reviewed.

8.2 STAR TOPOLOGIES

8.2.1 Single star

The simple structure of the single-star local network is shown in Fig. 8.1. With a dedicated link from exchange to customer, it is the simplest of all architectures, requiring only jointing points along the route. Traditionally in the copper network these have been above-ground flexibility points (e.g. 'cabinets'), where exchange-side pairs could be connected through to distribution-side pairs as and when appropriate. Such points are less desirable with a fibre network, where joints are lossy and time-consuming, and fibre organisation more awkward than with copper pairs. If the fibre network were being installed for general service to all types of customer, then the high take-up of telephony service nowadays would justify routeing lines all the way through to the DP (distribution point) near to the customer's premises. If it were being provided on a more selective basis, then a flexibility point seems essential to avoid running a new fibre all the way back to the exchange each time a new customer is added.

Where video services are provided these come from a cable TV headend. This could be co-located with an exchange, but more typically is further back in the network as shown in Fig. 8.1. The video services will sensibly pass via the exchange, where there is a cable TV hub-site; indeed in the single-star network this is where narrowband and broadband services will be combined to go on a dedicated link down to the customer.

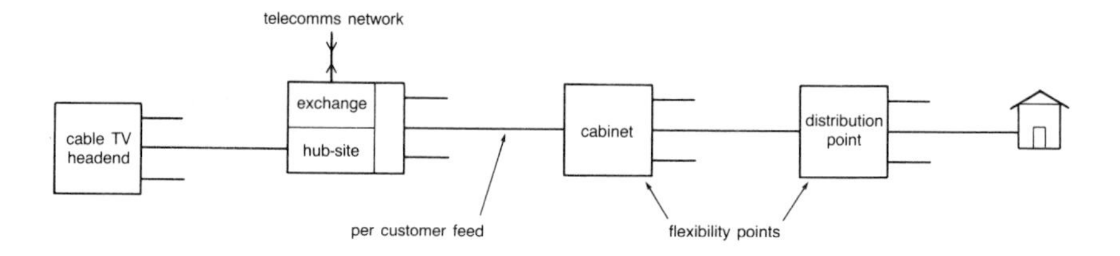

Fig. 8.1 The single-star network structure.

The characteristics of the single-star architecture for an optical local network can be summarized by the following advantages and disadvantages.

Advantages

- Simplicity — the dedicated fibre leads to the least restrictions in serving the customer, namely:

  — full fibre transmission capacity available to customer;
  — all line lengths in the local network are comfortably served by single-mode fibre;
  — technically undemanding on end equipment, low optical loss (compared to a splitter network) allowing the cheapest optical components.

- Full customer independence:

  — avoiding any interaction under fault conditions;
  — simple maintenance and testing;
  — straightforward network management;
  — ease of upgrade, affecting only the one customer.

- Benefit of centralized equipment — economies of scale in terms of maintenance, powering, racking, etc, at a single central point.

Disadvantages

- Cost — though simple, the provision to each customer of a full fibre link capable of vast capacity (beyond their likely demand) is wasteful. The cost of optical sources and receivers, fibre, cable handling and jointing is a significant portion of total cost per customer. At its purest form the single-star network gains no benefit from sharing of optical elements.

- Size — this is a fibre-rich topology and there is concern about the difficulties of handling perhaps tens of thousands of fibres into one exchange.

The evolutionary benefits of getting fibre all the way to the customer are attractive, but the challenge to do it using sensible engineering methods and at a reasonable cost is a hard one (particularly for the telephony-only customer). For this reason a single-star topology looks unlikely to be a general solution until optical component and handling prices have dropped dramatically. However where a customer has a high service demand (a large business) then single star is the sensible option; the cost of the dedicated feed is justified for present and future capacity and the simplicity, flexibility and security are highly desirable features. This has been the basis for much of the early provision of fibre direct to major business customers, e.g. the flexible access system (FAS) used by BT [1].

For the smaller business customer and the domestic residence other topologies need to be examined and the distributed star is one option.

8.2.2 Distributed star

The distributed-star structure is illustrated in Fig. 8.2. There is still a dedicated feed to each customer, but now it is a shorter link from a remotely situated access point (AP). Thus there are a number of 'mini-star' networks distributed throughout the local network. In the purest form of this structure each AP communicates back to the exchange (for telecomms) or headend (for cable TV) by dedicated (fibre) links; in American parlance this is often referred to as a 'star-star' or 'double-star' topology.

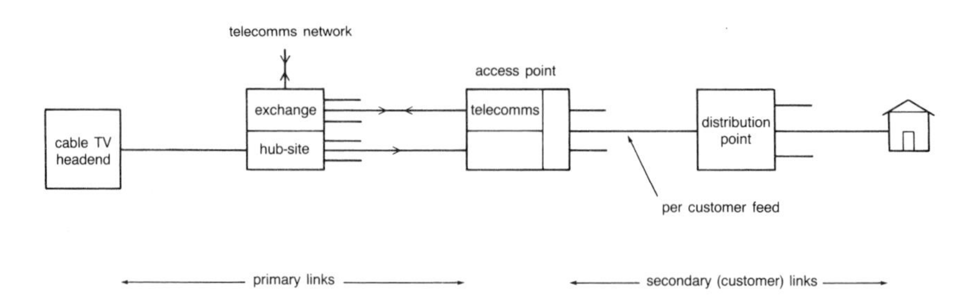

Fig. 8.2 Distributed-star network (with access point at cabinet position).

More usually though there is benefit in a combination of dedicated and shared links in this part of the network (termed the primary network/links), as shown in Fig. 8.2. Most obviously services specific to customers on the AP (e.g. a telephony multiplex, data links, video library channels) are carried on dedicated links, whilst broadcast services (e.g. entertainment TV) are distributed via a splitter network to many APs. However, economic and other factors may make it attractive to provide the dedicated services over shared links also; thus a multiplex is distributed amongst APs from which each takes a fixed, or dynamically changeable portion of the capacity (for example, the street-TPON scheme described in Chapter 9).

In comparison to the single-star network the distributed-star scheme has the following advantages and disadvantages.

Advantages

- From an evolutionary point of view it allows fibre to be taken closer to the customer, i.e. down to the AP at a cabinet or DP position, whilst still allowing the final feed to remain metallic. Metallic secondaries (copper pairs or small-bore coaxial) are an economic means of provision and for the foreseeable future place no limitation on service capability. If eventually fibre secondaries prove economically viable and otherwise desirable, then they can be installed as part of normal refurbishment.
- The network is lean in fibre, even if it is optical all the way to the customer. Multiplexing of telecommunications services and switching of broadband services at the AP reduce the fibre count dramatically on the primary side (even today one can conceive of just one broadband and one telecommunications fibre being sufficient for a cabinet serving several hundred customers). Fibre-handling problems are reduced along that route and back at the exchange or headend. If fibre goes to the customer there is still a handling problem within the AP because space is at a premium, but it has been broken down into a more manageable number of terminations.
- The customer link is now short, but the dedicated customer link has been maintained. It is therefore very undemanding technically, enabling low-cost solutions, and the single-star virtues of privacy, fault tolerance, and ease of maintenance have been retained.

Disadvantages

- Most obviously there is now street-sited equipment where before there was none. Housing equipment in a cabinet or underground requires more care in terms of environmental control, and can be difficult in terms of finding a site and arranging power feeds. It is often argued that this external equipment is a greater maintenance liability. However, experience in the field, as with the switched-star network (SSN) deployed in Westminster [2], has shown this not really to be the case as long as the environmental aspects within the housing are well designed. Indeed such equipment is probably less vulnerable than that in the customer's premises, and is certainly more accessible by maintenance personnel.
- The intrinsic flexibility and adaptability of the single star is somewhat compromised. The constraints on size of the AP housing, and to a lesser extent the deliberately limited fibre capacity on the primary side, may restrict the expansion capacity for future, perhaps unforeseen, enhancements.

Except for high-revenue customers (i.e. medium and large businesses) the distributed-star scheme must be favoured for the foreseeable future as the economic star structure.

8.2.3 Alternative structures

There are a number of variations possible on the distributed-star scheme and two are worth a brief mention (although it is arguable whether these hybrids can still reasonably be called star networks).

Firstly radio is an option as the means of final distribution, rather than cabled media, but is likely to operate in a broadcast mode. Entertainment TV fits naturally into this scenario with a transmitter on a pole by the AP and receivers on the subscribing homes (see Fig. 8.3). The short distances mean that frequencies as high as 30 GHz can be used. Two-way communication is more difficult, but can be achieved using TDMA techniques developed for multipoint radio systems (and adapted for TPON).

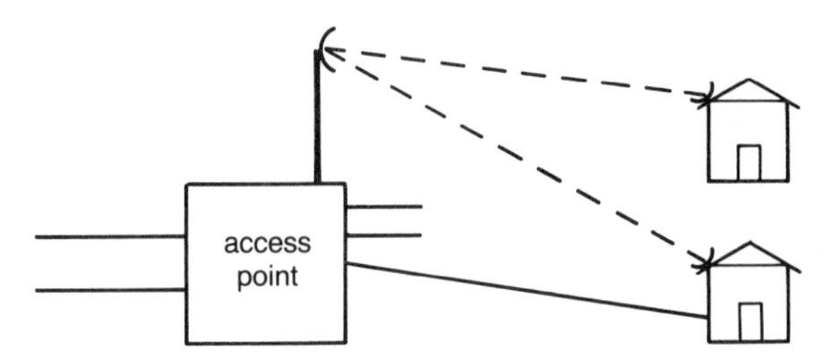

Fig. 8.3 Access point with radio and/or cable links.

A second approach is to still use fibre on the customer link, but to utilize some sharing there. Most obviously splitters (2-,4- or 8-way) are deployed at the DP (see Fig. 8.4). This is now a marrying of star principles with those of passive optical networks (see Chapters 9 and 10). The cost benefit comes both from sharing optical devices at the AP and the reduced fibre content; the smaller number of fibres to be handled within the AP is an added attraction.

The most straightforward sharing mechanism is to divide the fibre capacity into n fixed slots by electrical frequency division, optical wavelength or time multiplexing. The information that would have gone down the dedicated feed now occupies the customer's allocated slot. The alternative is to recognize that there is common information such as broadcast TV and audio which can occupy a commonly accessible slot, thus reducing the capacity required for each customer's dedicated slot. Still further variations are possible by

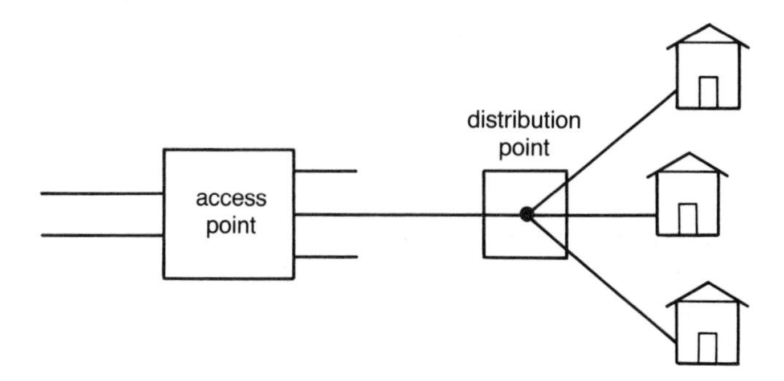

Fig. 8.4 Access point with fewer optically-split customer links.

dynamically allocating capacity to customers on the shared link as their demand varies.

There is though a balance to be struck between efficient usage of capacity and over-complexity.

8.2.4 Multiservice networks

The previous sections have discussed the general characteristics of a star-structured network, but in practice the strengths and weaknesses are also related to the services carried; a brief look at the main service types and their applicability to the star structure is therefore worthwhile.

Telephony — being a dedicated and interactive service to an individual it matches well to the dedicated feed of a star network (i.e. there is natural security and two-way operation is straightforward). If fibre is taken as the preferred medium to the home then a narrowband service like telephony occupies very little of the available bandwidth, and the dedicated star connection seems wasteful. Since very many customers in a multiservice local network will remain telephony-only for some time, the choice is between leaving the final feed as copper or devising a really cheap telephony-only fibre link. The latter may not have to quite match the low cost of a copper feed, since it provides a ready upgrade to other broadband services.

Data — the natural security of a dedicated feed is perhaps even more valuable for some types of data (e.g. financial information). However, data services are both sporadic, and mostly low bandwidth, so a fibre feed to the customer is hardly necessary for this service alone.

Broadcast TV — this is a non-specific service where the same information (i.e. a multiplex of channels) is required to go to a large number of customers. The most natural type of network to match this is one where an initial feed is split and fanned out as much as is compatible with retaining adequate performance quality. A star connection is also quite capable of carrying this multiplex; the question is how efficiently it does this.

The distributed star provides a compromise — fan-out occurs on the primary links and then channels selected at the AP are routed on to the customer's dedicated feed. One major advantage that accrues is that premium channels can be securely switched to those that have paid, rather than relying on sending scrambled signals into the home; this latter scheme is prone to programme theft from black market descrambler units. A second benefit is that, since the customer link only has to carry the channels requested by the customer at any one time, there can be access to an arbitrarily large, and expandable, number of channels at the input to the switch.

Dedicated video — an attractive service is a centralized video library from which a dedicated feed to the home can give the customer their own personally controlled video source for either entertainment or a more interactive purpose (e.g. learning). This has been operated in trial form within the Westminster SSN, where it proved very popular (as it always does in demonstrations). What remains to be shown is whether it is a viable commercial service. A fibre star-structured network clearly has the capacity, and the interactivity for this service. However, in the distributed-star case, there is the question of how many video channels to allocate on links between the library and the AP; clearly only a given percentage of customers will want this service at any one time, and occasional blocking will be acceptable, but the future level of take-up is hard to judge with such an untried service.

Videophone — a full two-way video service is readily carried on a star network since the dedicated fibre feed can have equal high capacity in both directions, allowing full bandwidth transmission. The advantage then is that relatively cheap end-equipment can be used. Concentration can occur at the AP by switching on to fewer primary links, and can again take place at the headend/exchange, after which video compression can be used to reduce the bit rate for long distance transmission.

However, if video compression eventually becomes cheap enough for use in the home then the balance of the argument may change. Dramatic compression to as low as 64 kbit/s is now possible for simple two-way low-definition video interaction, making this actually a narrowband service with the same characteristics as telephony (e.g. ISDN could carry it). The outcome of whether a high- or low-bandwidth approach will win out then hinges on the continuing debate as to whether the costs for transmission or for signal processing can be reduced most easily, and also on what videophone quality is required by the customer.

In summary the star-structured network can be seen to be appropriate for all types of service, both present and future. Its universal capability and enhancement potential are its great strengths. The question is whether it can provide this potential economically in an all-fibre scenario, particularly where many customers may only want narrowband services. This is the challenge being addressed by the BIDS network (broadband integrated distributed-star), which will be described in detail in section 8.4. Before this though section 8.3 will examine the main network features, and the technology now becoming available to allow this challenge to be met effectively.

8.3 KEY NETWORK FEATURES

8.3.1 Fibre optics

Optical transmission of digital signals has developed through trunk and junction applications to the state where there is a good range of equipment with steady advances in capacity (i.e. 560 Mbit/s moving through to 2.4 Gbit/s). For the higher bit rates at least (i.e. 34 Mbit/s and above) the standard technology used is a 1.3 μm laser with a PIN receiver of some type; thus the telecomms multiplex between the AP and exchange in the distributed-star network is readily catered for by available equipment.

The telecommunications link to the customer is something of a dilemma for the fibre optics designer in trying to match this high-bandwidth medium appropriately to such a low bit rate service. Digital operation suits fibre links, so it is attractive to base it on the 144 kbit/s ISDN standard to provide a future-proofed link (however, analogue possibilities should be regularly revisited as optical devices become more linear). The task is then a relatively undemanding one, and cheaper LED/PIN combinations are possible or perhaps mass-produced CD lasers. Multimode fibre would be quite appropriate for the distances involved, greatly easing jointing tolerances and bending loss problems, but single-mode fibre seems preferable to provide for enhancement beyond low bit rate telecommunications.

A single fibre to the customer is desirable to reduce fibre handling problems, and so two-way operation is required. A combined optical source/receiver/coupler in one package is an attractive solution to try to minimize costs, although a rather different and novel approach has been used in BIDS and this will be described later. With the very large volumes that the local network could produce, the low cost targets for these optical devices should be achievable; unfortunately manufacturers are reluctant to move in this direction before this volume market is more assured.

Video transmission is an interesting area where a variety of techniques vie with each other. With digital operation most natural for fibre, it might be expected that digital video would be dominant. However, so far it has been too expensive to compress a cable TV quality channel below the 50-70 Mbit/s region. Hence only as links at Gbit/s rates have become practical has it become possible to achieve reasonably large multiplexes of digital video.

Optimism is growing however that much larger compressions of the digital rate will become economically viable for per-customer equipment. This is largely based on the drive in the USA to find an HDTV approach compatible in bandwidth with existing broadcast signals, and also to find a means to dramatically increase the channel capacity of satellite transmissions to match cable TV. It remains to be seen how quickly this promise can be realized.

Except for high-quality or long-repeatered links, frequency modulation (FM) was favoured mostly during the 1980s over digital operation. The FM advantage factor helps overcome laser noise and nonlinearity, and modulators and demodulators are a mature technology and reasonably cheap. Until recently, capacity per fibre has been modest and similar to digital links (4-16 channels), but use of high-bandwidth lasers has allowed larger numbers of channels to be packed into the GHz region. Another recent advantage for FM operation is that it is the transmission format for direct broadcast by satellite to the home, and so matches to the cheap mass-produced set-top decoders now appearing for this service.

Early on in the development of video transmission over fibre, the use of amplitude modulation (AM) was rejected as far too demanding of the optical devices. This has now been turned round in the last two years by the arrival of lasers that are both highly linear and very low noise. Such devices are still very expensive, but this is compensated for by the modest bandwidth requirement of AM, allowing a large multiplex which most importantly matches to the existing carriage technique on a conventional cable TV network. For the next few years AM operation must remain a favoured option for delivery to the home since this is also the present standard TV input format. Additionally there are developments of lasers employing a quantum confinement structure which show that further substantial improvements in linearity, noise and efficiency will be achieved.

In giving the customer a dedicated fibre the star network makes the least demand on fibre capacity. If video channels are switched on to the customer link then two, three or four such channels may be required, and so cheap low-bandwidth technology can still be applied. On the primary links larger multiplexes are desirable, but of course these are less cost-critical. The need for enhancing fibre capacity with wavelength division multiplexing (WDM) appears small, certainly on the customer link, although it is possible that HDTV could eventually require this. More likely is that WDM will find application in one of the hybrid schemes mentioned earlier as one of the more secure and straightforward ways of sharing a link between a number of customers.

8.3.2 Switching

Telecommunications switching will be performed back at the exchange (or remote concentrators) in the normal way and is not part of the local network design being considered here; there will be telecommunications multiplexing at the AP and the techniques for this are covered later. However, this section will briefly look at the broadband switching required for the video signals at the AP (and perhaps used also at the headend).

The switching can basically be done in three ways — space, time, or frequency; space switching is most straightforward. For cable TV type service a broadcast switch is required where each input can feed any number from one to all the outputs; thus fan-out of the input is a crucial part of the design. In practice the AP may have a mix of broadcast feeds and dedicated feeds (e.g. for video library). In the Westminster SSN this led to a two stage set-up where broadcast channels went direct to the final stage and dedicated channels were shared via an initial switching stage. Only the latter could suffer from blocking under overload conditions.

The technology used for the crosspoints depends somewhat on the form of the signals. It is sensible for the switch to match to the transmission techniques on either side of it. For analogue operation suitable CMOS and DMOS FET crosspoints are available with modest integration levels within a chip. Translating the video signal down to baseband (0-6 MHz for PAL) makes the design of both switching and buffering for fan-out easier, but higher frequencies can be catered for, allowing signals on a carrier to be switched.

Such analogue crosspoints can also pass digital signals, but in this case it is really more sensible to utilize digital gates, where much higher levels of integration are achievable and some regeneration of the signal effectively occurs. ECL matrices are the most readily available, but take a lot of power. Very impressive CMOS arrays have recently been developed using small device geometry, enabling them to pass 200 Mbit/s with low power demands.

These digital arrays have generally been used as 'slow' space switches, i.e. the signal itself is fast but the switching rate is slow. The alternative is to employ time switching, perhaps allied with space switching. Thus a space switch may first choose from a number of incoming digital multiplexes, and then gate out the required channel from that multiplex and delay it to fit the desired timeslot on the output (as illustrated in Fig. 8.5).

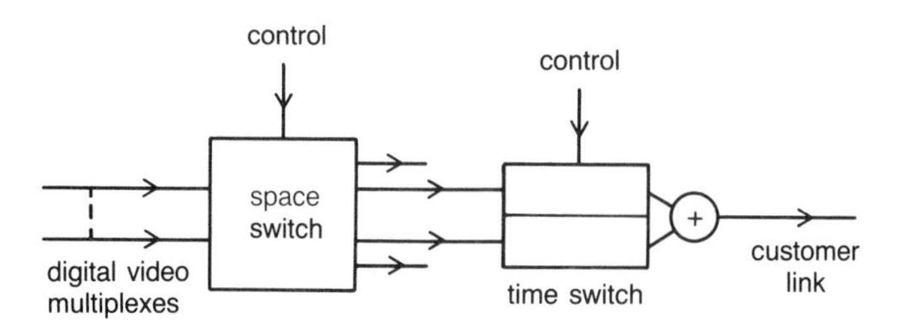

Fig. 8.5 Switching arrangement to select two digital video channels per customer.

Analgous to time switching of digital signals is frequency switching of analogue signals (used occasionally in coaxial cable TV schemes). Again an incoming multiplex of channels is selected and the chosen channel translated into the required bandwidth slot on the output by mixing with a signal from a frequency-agile oscillator (Fig. 8.6). A bandpass filter then ensures that only that channel goes through. The channel selection then is achieved by an appropriate change of the oscillator frequency.

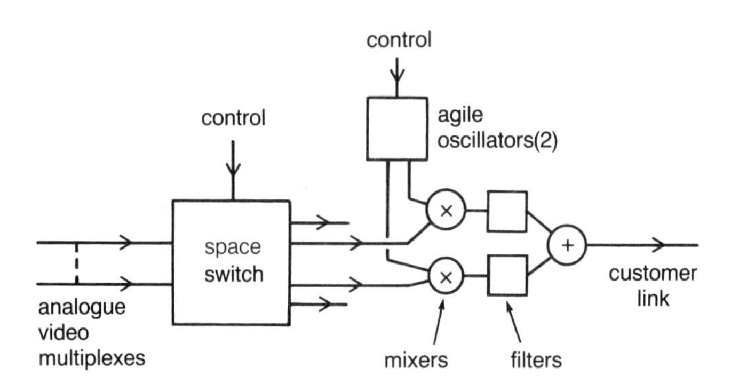

Fig. 8.6 Switching arrangement to select two analogue video channels per customer.

Optical switching is yet another possibility with a variety of techniques and technologies too numerous to describe here. It will probably start to be used in a small specialist way, such as to switch to alternative fibre routeings, but is unlikely to replace electrical switching entirely even in the long term.

8.3.3 Cabinet engineering

The distributed-star topology requires a considerable amount of complex electronic equipment, under local microprocessor control, to be located at each AP. The housing for this equipment is therefore of prime importance, especially since a grouping of 200-500 customers is considered the economic AP size. This currently implies an equipment volume of about one cubic metre; too small for a purpose-built building at each location.

One alternative is to rent space in an existing building, and in some cases this has proved to be a practical solution. However, experience has shown that it is by no means a general one, raising such problems as access and security. A solution which can work well is to build environmentally controlled below-ground chambers, along the lines of a large deluxe footway box. The American experience of this method of equipment housing is good, but the civil engineering costs for a UK application are prohibitive. This leaves the roadside cabinet as the favoured solution. Since the required cabinet is somewhat larger than existing street furniture, the problem is one of finding suitable sites at the required locations.

The cabinet design is no simple task. It must provide a physically secure housing and an environment suitable for the operation of complex opto-electronic and microprocessor equipment. Glass-reinforced plastic (GRP) and stainless steel have been used to fabricate two different designs for assessment and comparison. Both types are used, in significant numbers, in the Westminster SSN. The stainless steel version has the edge so far, with advantages in handling and radio frequency (RF) screening.

Temperature and humidity are the two most important factors, and these are controlled by use of a closed-loop heat exchanger and an air-sealed cabinet. Figure 8.7 shows the basic principles of the cabinet thermal management. With such a high equipment density in a confined space, forced air fan cooling is a necessity. A source of heat input, secondary to that generated by the equipment but still significant, is solar gain. In order to minimize this effect, the shell construction is double-skinned, with an interskin foam insulant. The cabinet is alarmed for out-of-specification temperature and humidity, with automatic feedback to the headend control room.

Radiation ingress and egress are important factors, with about 30 dB of RF attenuation in the band up to 1 GHz provided by the cabinet. The electrical integrity of the shell required to achieve this is maintained by use of composite weatherseal and RF gaskets on all doors and construction joins.

In the SSN and BIDS designs, the equipment racking within the cabinet is based on standard 19 inch rack practice with mostly double Eurocard shelves. This construction gives an economic solution. By way of example the layout in the latest Mark 2 SSN design is shown in Fig. 8.8.

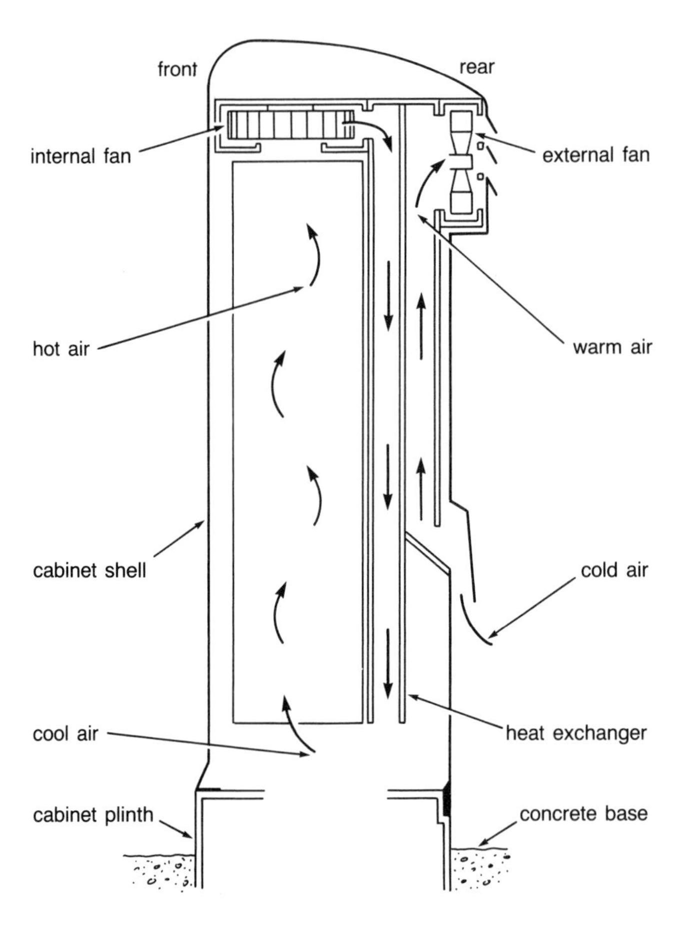

Fig. 8.7 Cabinet thermal arrangement.

Maintenance of equipment in these street cabinets is of prime importance and, since most are sited against a wall or fence, total front access is desirable. To achieve this the latest design has only shelf backplanes; no rack backplanes are used and all shelf input/output ports are brought to the shelf front by interconnect cards. Interconnection between shelves is then made by cable, as can be seen in the photograph of the Mark 2 SSN cabinet in Fig. 8.9.

The cabinet acts as a flexibility point. Thus in the SSN design some 300 dedicated customer coaxial tubes are terminated on a patch panel in the cabinet. Pre-terminated tails are used; these are spliced on to the route cable in a footway box associated with the cabinet. Each new customer is allocated a customer service module (CSM) which is the unit taking outputs from the switch and transmitting the required service package over the customer link.

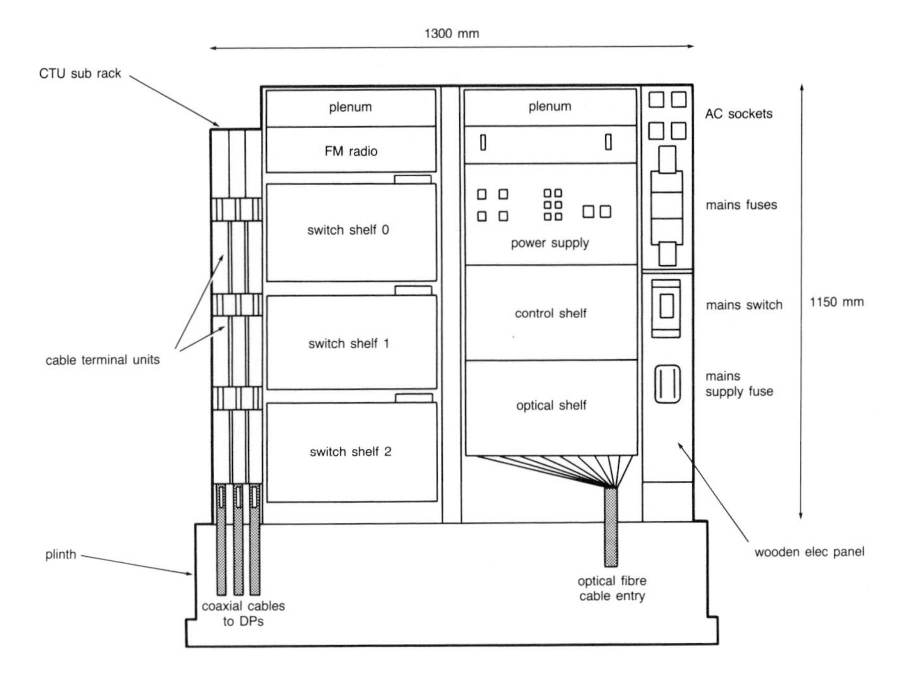

Fig. 8.8 Internal cabinet layout of the SSN Mark 2.

The allocated CSM is patched across to the correct outgoing coaxial tube on the patch panel.

Where fibre is to be used on the AP-to-customer link, the arrangement is somewhat different. Fibre patch cords are to be avoided if possible and hence a fibre management system, allowing direct termination on to the optical CSM, has been developed for BIDS. The customer fibre can be installed by blown or conventional fibre cable techniques, but in either case the terminating ends are fed into the cabinet and on to trays in a fibre-management unit such as that seen in Fig. 8.10. Each tray in the unit accommodates up to five fibres. The fibre ends are fed out from the tray and each is fitted with a field terminable connector, as part of the customer installation, when service is requested. Connection is then made directly to the appropriate CSM optical output.

Since the links to the cabinet are optical fibre, there is no convenient opportunity for central power feeding to the AP. Local mains power is therefore used via a circuit breaker and fuses. Where telephony service is provided, back-up power in the event of mains failure is needed.

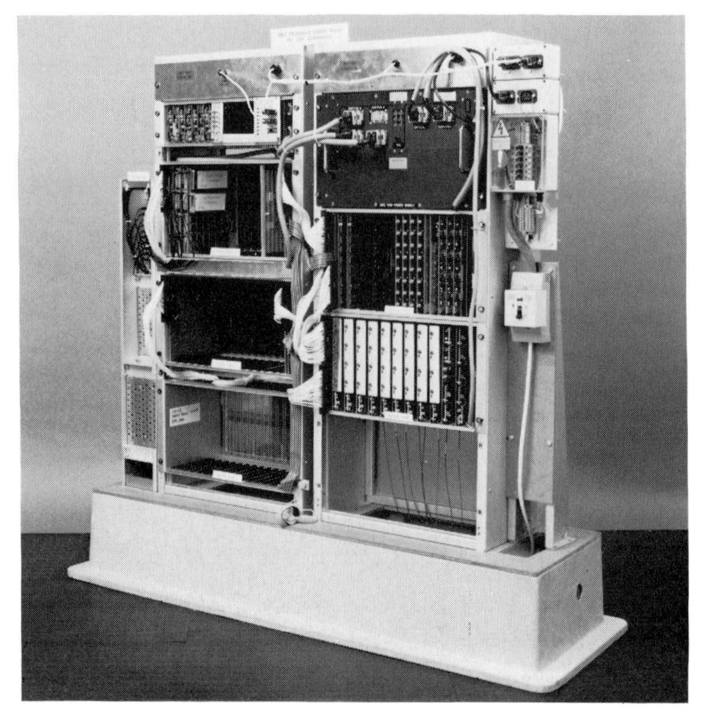

Fig. 8.9 Internal cabinet racking (partially equipped) for SSN Mark 2.

For a fully equipped AP, standby power of some 500-600 W for a period of four hours is required. Sealed lead-acid battery technology is adequate to provide such an energy store, utilizing the space available in the cabinet plinth or the adjacent footway box. Standby power also requires a.c./d.c. and d.c./d.c. power supplies, together with control and monitoring circuitry.

8.3.4 Customer equipment and home distribution

This is a large area of study, and an important one for the success of future multiservice networks. Only a few of the major issues can be touched on here however.

The star network leads to relatively simple customer equipment, since the service selection is performed back in the network. Basic telephony has a straightforward interface though there can be problems ensuring that the full variety of equipment nowadays connectable to a telephone socket can be catered for. Where telephony is particularly demanding is in the required high availability of the service, in particular the need to continue in the presence of power failure. Fortunately advances in battery technology are

Fig. 8.10 Fibre-management unit.

now being made that will help, e.g. lithium-polymer rechargeable units with enhanced capacity over nickel-cadmium ones.

Traditionally cable TV has used a separate set-top unit to select the required channel and translate to the TV set's UHF input. Although 'cable ready' TV sets are available in the USA, the separate set-top unit has proved to be an acceptable solution for customers, and, if well designed, can even enhance the customer's perception of the system. Of similar importance is the customer control mechanism via a built-in or infra-red keypad; this must allow easy use of the services offered. Text displays can guide the customer through complex procedures (e.g. setting up the system to provide channels at appropriate times for VCR recording). Text generation is thus another essential technology; the display can be generated from data by centralized units in the AP, or even headend, but there are increasing arguments for a text generator in the customer's equipment (the teletext facilities in some sets can also be used).

In a multiservice situation the delivery points can be spread around the home. The telephone socket is usually acceptable close to the entry point into the home, where the network interfacing circuitry will also be located. Set-top units will be required at one or more points for TV service; audio is usually provided at the same points, traditionally in FM radio format, but baseband or digital stereo signals may be preferred soon and may be required in a different place from the TV. Data may go to the personal computer in yet another room. Much attention is therefore being given to home distribution systems, but not always with video in mind. The problem is to decide the best format to carry video round the home — should it be the standard AM over coaxial cable or the more robust FM or digital forms, and is fibre a serious proposition?

Finally safety must be mentioned as an important consideration. Optical safety may or may not be a concern depending on device powers. Isolation between the network and customer-owned equipment is the other main issue, whereby damaging voltages from either must be prevented from crossing the interface. Fibre links provide intrinsic isolation for most of the network, although there is still the network terminating equipment on the customer's premises.

8.4 THE BIDS NETWORK

The BIDS network was designed as a practical realization of a broadband, multiservice, all-fibre system for the short to medium term, and as such has been deployed in the local access trials at Bishop's Stortford [3]. The design is essentially a development from the existing MkII version of the Westminster SSN, providing broadband, combined with remote multiplexer equipment for telephony and data service. It therefore combines the SSN analogue transmission for broadband services, with digital transmission for telephony and data. BIDS is also intended to be operated by extensive system administration and network management software, which has been further developed from that produced for the SSN.

8.4.1 System description

BIDS in its full form has the typical distributed-star architecture shown in Fig. 8.11. Broadband programme material from terrestrial and satellite broadcast services, video tape and local studio are assembled at the headend. The headend also houses the network control and monitoring facilities, and the necessary gateways and servers for interactive videotex services.

Signals are then transmitted to a number of hub-sites on 1.3 μm single-mode fibre primary links. Each fibre has a capacity for 16 broadband channels in an analogue FM transmission format. The channel spacing and bandwidth, as for all links in the network, are designed to accommodate PAL or DMAC TV signals.

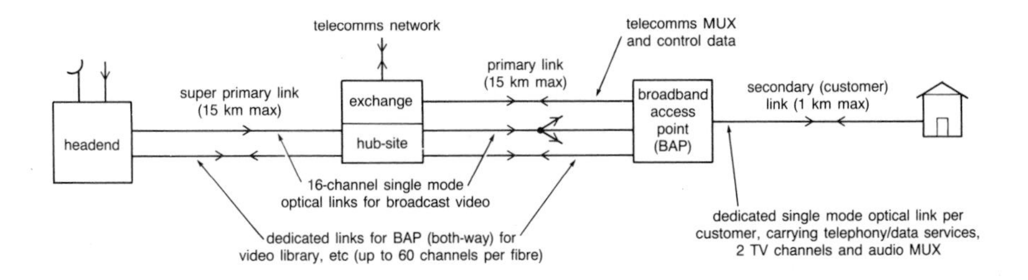

Fig. 8.11 The broadband integrated distributed-star network.

Hub-sites are intended to be located at local telephone exchanges and are therefore also the point at which BIDS interfaces to the telephony network. In small systems, and as implemented for the Bishop's Stortford trial, both headend and hub-site can be co-located at the local exchange. In the trial system modified remote telecommunications multiplexer equipment, based on the BT LA30 design, is used to effect the interface to the local System X exchange, supporting the 30-channel analogue DASS II software.

From the hub-site broadband transmission down to the cabinet is via primary optical links identical to those described above. This cabinet is known as the broadband access point (BAP) in BIDS. Transmission of telephony service is over a 140 Mbit/s synchronous multiplex link in both directions. The BAP is the major functional element of BIDS and is shown diagramatically in Fig. 8.12.

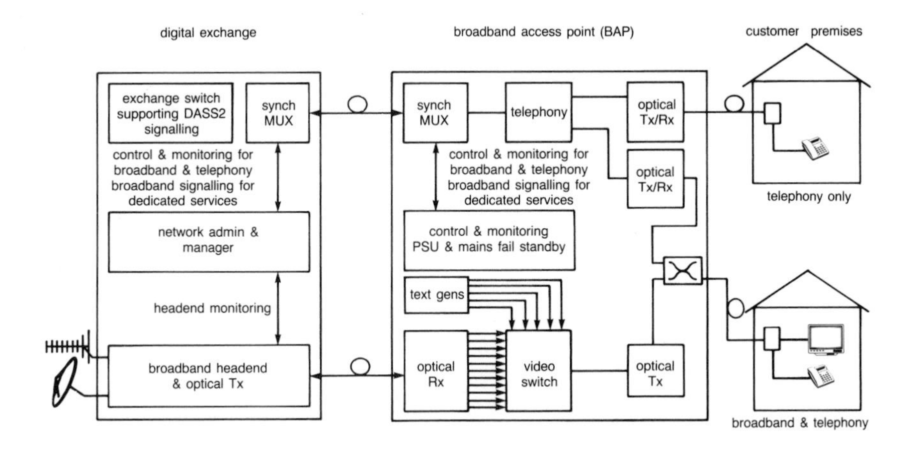

Fig. 8.12 The main functions of the broadband access point.

The incoming broadband channels are individually translated down to a common 25 MHz centre frequency prior to switching, retaining their FM format. The broadband analogue switch has a non-blocking, 48-input channel capacity. Accepting a degree of blocking, for minority interest programmes, allows the total number of input channels to be considerably increased by the introduction of a second stage switch. Each customer on the BAP has two dedicated switch outputs, whose signals are translated up to centre frequencies of 175 MHz and 225 MHz. These two channels are then passed to the broadband laser drive for transmission over the final link. The total spectrum on this link is shown in Fig. 8.13.

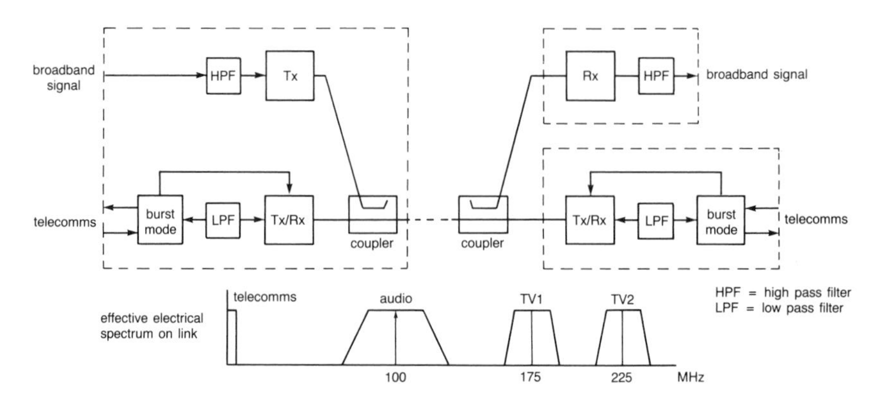

Fig. 8.13 The BIDS optical customer link.

From the telecommunications primary link 24 of the 2 Mbit/s data streams are processed by the burst communication circuitry; each stream serves ten customers, thus providing for a total of 240 customers. The generous capacity available in the 140 Mbit/s link is used to avoid the need for conversion to common channel signalling at the BAP. By allocating three 64 kbit/s slots per customer, channel-associated signalling can be retained, eliminating conversion circuitry and thus saving valuable space and power load in the cabinet. Power load is a significant factor because, for telephony, the BAP must be provided with mains-failure standby facilities as discussed earlier.

An essential requirement for BIDS is a low-cost, telephony-only customer link; to achieve this a novel approach using burst mode operation has been developed. Alternating compressed bursts of downstream and upstream digits allow the continuous two-way signals to be conveyed over the single customer fibre using simplified optics (because transmission is only in one direction at any one time). Indeed one solution is to use a laser both as a normal transmitter to send a burst and then, removing its bias, to let it act as a receiver for the incoming burst. Alternatively, as used in the Bishop's Stortford trial, transceivers can be utilized with greatly simplified coupling between the fibre and the transmitter and receiver devices. Figure 8.13 shows the complete optical secondary link, illustrating how the broadband and telecommunications signals are optically coupled together and, although working at the same optical wavelength, are separated out by electrical filtering.

Figure 8.13 also shows a block of hi-fi audio channels sent on the customer link, centred at 100 MHz. This is a digital multiplex of 16 stereo channels, encoded in a fashion akin to compact disc techniques. It is sent over one of the primary links to the BAP and then distributed to all customer links.

In the customer's premises, network interface equipment is needed which provides facilities for:

- the opto-electronic interface,
- telephone service in standard format,
- broadband service in standard PAL or DMAC format to a UHF TV receiver and in baseband format with stereo TV sound channels,
- hi-fi audio service and stereo TV sound channels, as pre-amplified left and right audio channels,
- power supplies, including standby power.

All broadband services, including videotex and video library, are controlled via a set-top unit with integral key pad, or by an infra-red remote control key pad. Signalling for the broadband services is transmitted via the telephony link to the BAP, where it is extracted to operate the switch. Broadband signalling which requires access to headend equipment, e.g. for videotex, is re-inserted into the primary telephony link for onward transmission. Videotex information is transmitted from headend to BAP as digital data. At the BAP text generators convert the data to analogue FM TV format for transmission to the customer.

8.4.2 The coaxial customer link option

The BIDS development to date has focused on early implementation for the Bishop's Stortford trial and has therefore fully embraced the 'fibre to the home' principle. This concept is seen to have significant advantages for long-term costs and for provision of more advanced services, i.e. it gives a better degree of future-proofing for the network.

However, in the short to medium term, it is difficult to see the costs of the customer link opto-electronic equipment falling to economic levels, especially for telephony-only service. For this reason variations of BIDS, using copper for the final link from BAP to customer, should be considered.

Perhaps the most obvious route to a viable network for early deployment, is to revert to small-bore coaxial customer links. Available bandwidth on such a link is ample to carry the multiplex of TV, radio and telephony services initially intended. The circuitry for both the broadband switching and transmission and the telephony burst communications circuit remains unchanged. It is necessary only to remove the optical transmitters and insert a simple electrical multiplex with a launch power amplifier.

8.4.3 Future development

It is quite probable that TV programme material will be produced, stored and transmitted in digital format. The domestic TV receiver will also become more digital in operation, and have a digital input.

At that time current BIDS broadband analogue transmission and switching would clearly be inappropriate. Consequently, future development of the BIDS system includes modification of primary and customer links for digital transmission, and use of a digital switch. Suitable transmission and switching techniques are already being studied for the BPON scheme (see Chapter 10).

The change to digital working will be driven by cost advantages and therefore awaits the advent of digital TV. However, once the change is made, there is considerable scope for further integration of all services, perhaps multiplexing the telephony and broadband together.

The use of existing multiplex equipment for BIDS telephony service was always seen as an expedient to enable early deployment. Perhaps a more elegant and economic solution is to use a street-sited multiplex version of the telephony on passive optical network (STPON), currently being installed as part of the Bishop's Stortford trial. Telephony and broadband signals would then be integrated on to a single coaxial link to the customer.

8.5 THE FUTURE FOR STAR-STRUCTURED NETWORKS

There is no doubt that star-structured optical networks are one of the main contenders for the future multiservice local networks being considered around the world. They can meet all the likely demands of future services with the minimum of constraints (particularly in the single-star case). Section 8.2 briefly went through the main existing service types and their characteristics as far as star networks were concerned. It is worth also considering two major developments for the future, HDTV and ATM, to ensure that these can be adequately catered for.

HDTV basically implies an increased capacity to be carried by the network. In a compressed form (e.g. the European HDMAC) it may require capacity of only 140 Mbit/s (digital) or 16 MHz (analogue). In this case the network impact is small, and there are now proposals for much greater compression, approaching conventional TV channel bandwidth. Even a fuller high-definition signal, requiring between 280 Mbit/s and 1 Gbit/s, or at least a 30 MHz analogue bandwidth, could be accommodated comfortably on a star network with an optical link to the home. Only if several such channel slots are needed would a second optical wavelength be needed. A more awkward problem might be switching HDTV at a different bit rate or bandwidth to the standard channels, in that this implies a separate switching matrix with different characteristics to the main one. However the number of HDTV channels will be likely to be relatively small so, if required at all, the switch size will be modest.

Asynchronous transfer mode (ATM) is the name now most commonly used for the fixed packet length scheme foreseen as the potential universal means of transmitting all service types (see Chapter 12). It is most obviously suited to variable-rate services, which can include certain techniques of compressed speech and video. There are mixed opinions as to whether full bandwidth video is sensibly coded into packet form, because of the problems of processing such high-rate services. Whatever the scenario though, star networks are well placed to cater for ATM. One very likely hybrid scheme involves an ATM channel (say 150 Mbit/s maximum) being allocated on the customer link to cater for all services except perhaps entertainment TV. At the AP packet multiplexing occurs so that the traffic from all the customers can be efficiently transmitted back to the exchange/headend where packet routeing is performed. On the customer link the ATM channel is multiplexed with others of similar rate carrying the entertainment TV.

Thus there are no barriers to optical star-structured networks efficiently carrying all likely future services. This stems principally from the generous allocation of a fibre per customer, but the consequent challenge emphasized throughout this paper is to provide this economically for a range of service offerings from narrowband to broadband. The BIDS network is one of the first steps to achieve this, but it is recognized that there are further developments required.

The distributed-star structure is seen as an economic deployment strategy for now. Remote telephony multiplexers (fibre-fed, but with copper links to the home) are now with us in many parts of the world and their numbers are likely to increase significantly in the coming years. Thus a distributed-star structure is being laid down already with the potential to augment to a multiservice scenario. It is likely that only broadband services will justify fibre upgrade on the final feed to the customer in the near term, but if a telephony-only fibre link can approach copper costs, and there is confidence in an eventual high broadband take-up, then fibre to the home could become an accepted policy for all new installations.

In the long term the single-star topology could return to the fore if great strides are made in fibre technology (particularly its handling), and the passive optical network topology, sharing capacity via splitters, is already acknowledged as a major option for local loop deployment. In practice each may find its market application, and there may well be hybrids between the various network types. However, for the foreseeable future the distributed-star topology will remain central to multiservice local network development, and will be the straightforward option against which other schemes will need to be judged.

REFERENCES

1. Dufour I G: 'Flexible access systems', Int Symp on Subscriber Loops and Services, Boston, USA (September 1988).
2. Ritchie W K and Seacombe R: 'The Westminster multiservice cable TV network — experience and future developments', 15th Int TV Symposium, Montreux (June 1987).
3. Boswell E J and Kerrison A H: 'Initial experience with the Bishop's Stortford fibre-to-the-home trial', 17th Int TV Symposium, Montreux (June 1991).

9

THE PROVISION OF TELEPHONY OVER PASSIVE OPTICAL NETWORKS

C E Hoppitt and D E A Clarke

9.1 INTRODUCTION

The application of optical technology to the local network has been a field of increasing interest and debate over the past few years [1]. While it is usually assumed that optical networks will penetrate to small businesses and residential customers to provide an integrated range of services, it is recognized that the economics of both the technology and the revenue from new services are uncertain.

For telephone companies an entry strategy is required that will enable optical technology to be installed in a way that is reasonably economic for telephony, but readily allows upgrade to carry the new broadband services when there is an opportunity for profitable investment.

One approach to fibre provision is a direct translation of the copper network based on a single star; this approach has been successfully used in BT's FAS (flexible access system [2]), aimed at the large business user, where the high cost of optical components can be shared across many 64 kbit/s circuits. Such cost sharing is not so effective for the smaller business and not possible for the single-line residential user. Several proposals [3,4] have been made based upon the use of optical links to electronics at nodes in the local access network with the final link completed in copper pairs. This shares the optical costs over the customers serviced from the node but does not address the need to deliver fibre to the home in preparation for the provision of broadband services.

Recently, BT has been studying methods for fibre deployment based upon passive optical networks [5,6] such that initial deployment can be tailored to telephony users, yet an upgrade is still possible as demand for broadband services matures. This new approach is called TPON (telephony over a passive optical network) which is being trialled at Bishops Stortford in the UK [7]. As a result of the success of that trial, other countries, including the USA, Japan, Germany, and the Netherlands have now started similar trials.

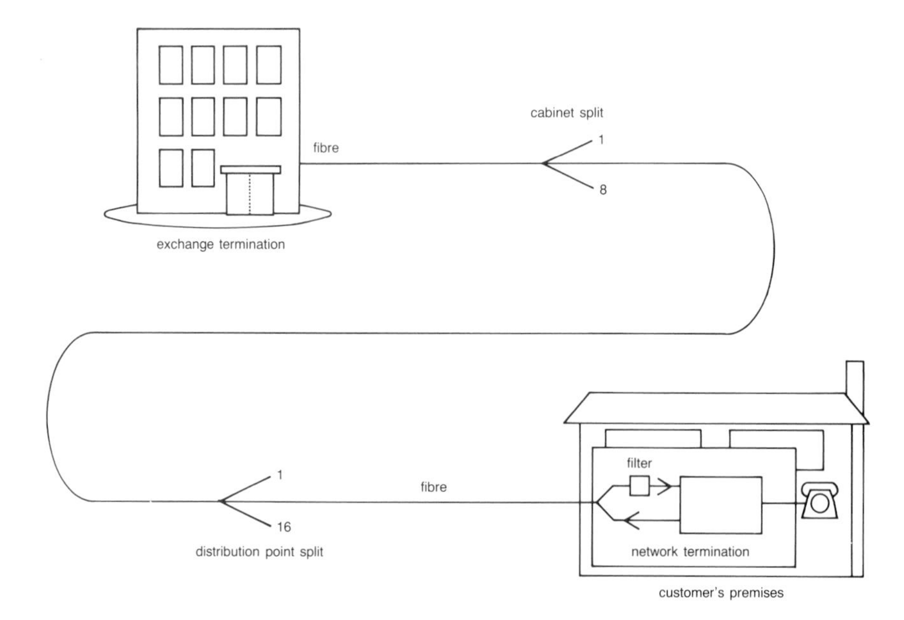

Fig 9.1 The TPON network.

The basic network is shown in Fig. 9.1. A single fibre is fed from the exchange, and fanned out via passive optical splitters at the cabinet and distribution point (DP) positions to feed a number of individual customers. A time division multiplexed (TDM) signal is broadcast to all terminals from the exchange on a single optical wavelength. After detection by an optical receiver, the terminal equipment selects the channels intended for that destination. In the return direction, data from the customer's terminal is inserted at a predetermined time in the TDM frame to arrive at the exchange within the correctly assigned time slot. The time management of the system is controlled by a ranging protocol that periodically determines the path delay between the network termination and the exchange and updates a programmable digital delay element in the network termination. An important feature is the inclusion of an optical filter in the network termination equipment that passes only the TPON system wavelength to allow other wavelengths to be added at a later date for new services without disturbing existing receivers.

This chapter describes the system options for such a network and discusses the design processes behind the construction of a laboratory demonstrator for a TDM/TDMA system based upon a 128-way split. The chapter also includes comments on how such a system could be deployed in the field, and the opportunities as well as the problems for network management and maintenance.

9.2 OVERVIEW OF SYSTEM OPERATION

Before looking in detail at the system options it is useful to consider a specific example of the way in which a telephone can be connected using TPON (Fig. 9.2), and in particular the role of the TPON bit transport system (BTS). The detail behind the choice of system parameters is covered later.

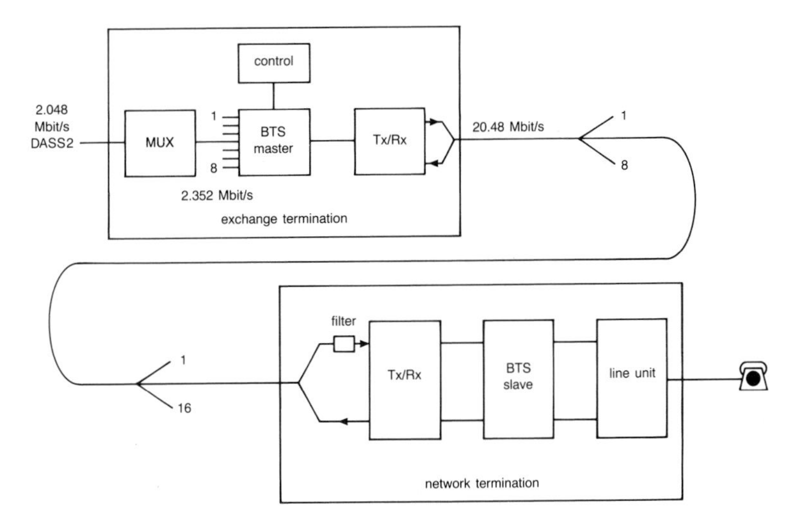

Fig. 9.2 The TPON bit transport system.

Speech channels emerge from the exchange as part of a 2 Mbit/s PCM stream with BT's digital access signalling system 2 (DASS2) (see Chapter 13). Within the TPON exchange termination, the statistically multiplexed signalling for the 30 speech channels in timeslot 16 of the DASS2 stream is converted to a form which can be delivered to individual network terminations. The signalling for each speech channel, formatted into 8 kbit/s channels is used, for example, to control ringing and to provide loop disconnect detection. The 64 kbit/s speech channel and its associated 8 kbit/s signalling channel are then multiplexed with the other 29 channels to form a 2.16 Mbit/s ($= 30 \times (64 + 8)$ kbit/s) signal which, combined with some spare bits, forms the 2.352 Mbit/s interface to the BTS hardware. The combination of speech signals and associated signalling will be referred to as 'traffic data'. The function of the BTS is to transport traffic data transparently to and from the network termination. Within the BTS-Master the 2.352 Mbit/s stream is bit-interleaved with similar streams derived from up to seven other PCM/DASS2 streams, giving a total traffic data bit rate of 18.816 Mbit/s which, when control overheads are added, becomes 20.48 Mbit/s on the optical network.

The passive optical network intrinsically ensures that all of the 20.48 Mbit/s stream is received at all of the network terminations. Each particular network termination can be instructed by the BTS-Master to select its traffic, starting anywhere in the multiplex, and extract a given number of 8 kbit/s channels that were contiguous in the original 2.352 Mbit/s streams. In the case of a speech channel the network termination selects nine 8 kbit/s channels, 64 kbit/s for speech and 8 kbit/s for the associated signalling. The line unit then converts these to a form suitable for the connection of an ordinary telephone.

In the reverse direction a reciprocal process occurs. The $64 + 8$ kbit/s traffic data from each terminal, plus certain control signals, are injected into the optical network at a predetermined time such that the signals from all terminals appear as a continuous and perfectly interleaved stream at the exchange terminal. The 'ranging' mechanism which ensures that this happens is controlled by the BTS-Master using the control signals in the 20.48 Mbit/s multiplex.

9.3 SYSTEM OPTIONS

9.3.1 Configurations

The simple approach above can be extended to take into account a mix of customer's telecommunications requirements. In particular, small business customers wanting, for example, five lines need not be served via five network terminations; instead a single optical transmitter/receiver can be used to serve a number of line interface circuits (Fig. 9.3). As well as being physically smaller this configuration makes more cost-effective use of the optical components by sharing them across several circuits. For convenience in terminology, this configuration is referred to as 'business TPON' to distinguish it from the single line 'house TPON'. Since all traffic data is available to all network terminations, a business TPON termination can access circuits from different 2.352 Mbit/s exchange termination ports — for example one port may carry telephony circuits, whilst another could provide private circuits or ISDN. Not only can the BTS inherently provide this 'grooming' function, it can also ensure that each of the 2.352 Mbit/s inputs is fully loaded before calling into service another exchange port. This is referred to as 'consolidation'. TPON's ability to groom and consolidate for the small business user increases its cost effectiveness in comparison with dedicated (and perhaps part-utilized) standard 2 Mbit/s circuits for each type of service.

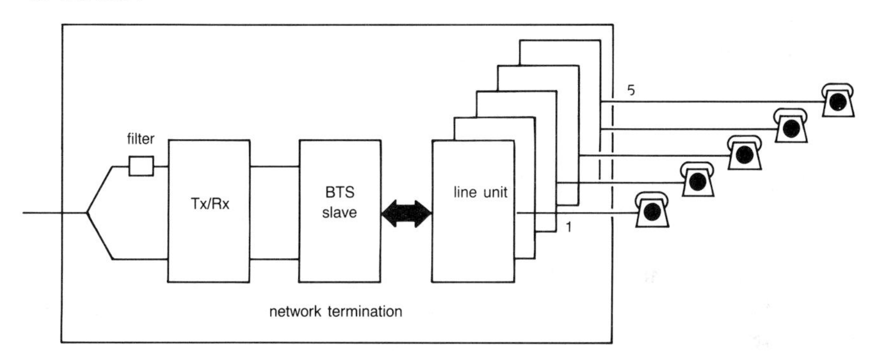

Fig. 9.3 Five lines business network termination.

A further variant is referred to as 'street TPON' (Fig. 9.4). This is similar to business TPON in that it is a multiline unit, except that it is situated in street furniture or in a footway box. The final link to the customers' premises retains the copper pairs. Because of the sharing of the fibre network and optoelectronics, street TPON is potentially more cost-effective than conventional street multiplexers and can be used as an early means of deploying fibre in the local network. Later, as costs fall or broadband services are required, the multiplexer can be removed and fibre links taken directly to customers in the form of house TPON; no further changes are required to the optical network or exchange termination for the telephony provision. Hence a major advantage of TPON is its ability for a single network to mix and match between residential and business customers as shown in Fig. 9.5.

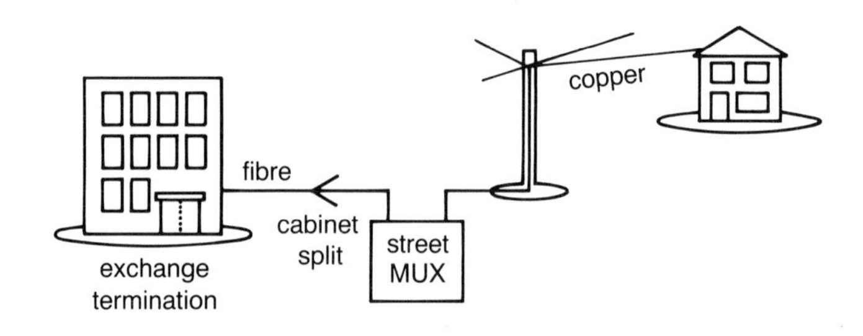

Fig. 9.4 Street TPON.

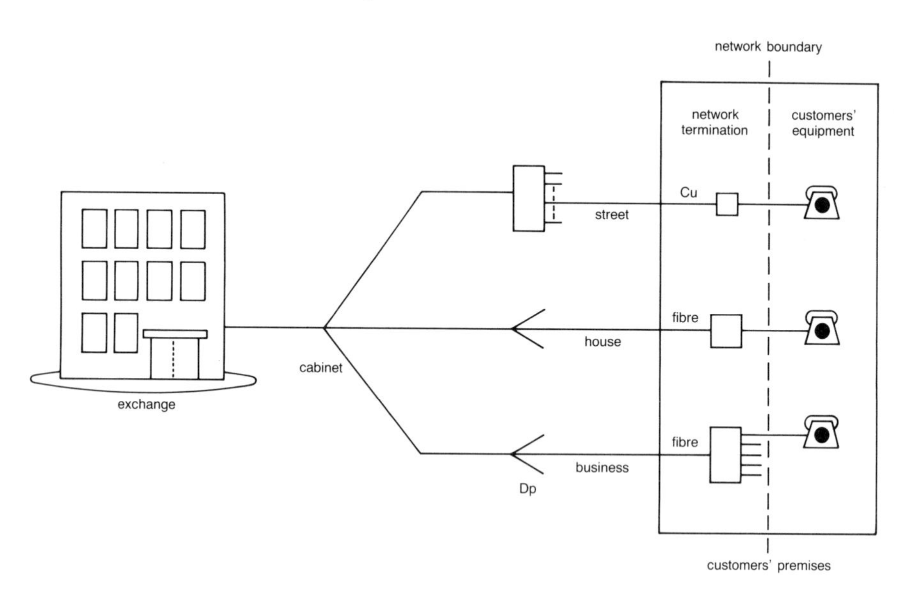

Fig. 9.5 Network termination options.

Recently a whole new set of nomenclature has arisen in the international debate on fibre access — fibre-to-the-home (FTTH), fibre-to-the-curb (FTTC) and fibre-to-the-office (FTTO). House TPON is an example of FTTH, business TPON an example of FTTO, and street TPON an example of FTTC. Also the network termination is often referred to as the optical network unit (ONU). Currently BT is playing a leading part in the standardization process for optical local access networks. In the USA, the major emphasis is on fibre-to-the-curb, while in the UK fibre-to-the-office for small and medium businesses is seen as the priority. The different approach is partly due to the different regulatory regimes in the two countries.

9.3.2 Choice of network bit rate

There are many factors which affect the choice of bit rate transmitted over the fibre. A higher bit rate leads to many advantages:

- larger number of channels available for sharing and circuit management;
- larger number of customers can be connected;
- better economics from a higher degree of sharing of expensive components and fibre.

Set against this are technology breaks — for example the current practical limit for low-power operation of standard CMOS is in the tens of Mbit/s. There are also practical limitations on the optical splitting ratio, and on the complexity of the ranging process. It is the authors' view that bit rates in the region of 10 to 30 Mbit/s offer the best compromise in the medium term.

The system bandwidth is used for the following purposes:

- traffic (e.g. 56 or 64 kbit/s for speech);
- call supervision signalling to the line units;
- housekeeping to control the operation of the BTS, laser power level and multiplexing;
- ranging control during quiet time;
- synchronization (from exchange to network termination);
- maintenance features such as fibre identification.

The lowest common multiple of all narrowband services was identified as 8 kbit/s and this was chosen as the traffic and signalling bandwidth granularity. Thus, for example, a speech service is offered using a minimum of nine 8 kbit/s timeslots, eight for speech and at least one for the channel-associated signalling; likewise ISDN can be offered using a minimum of 19 timeslots, 18 for 2B + D and at least one for conveying the layer 2 primitives. This 8 kbit/s granularity extends to the BTS housekeeping channel which is a dedicated 8 kbit/s control link to each optical termination on the network. Details of the BTS frame structure are given in section 9.5.4 below. The 8 kbit/s rather than 64 kbit/s and 1.544 Mbit/s rather than 2.048 Mbit/s circuits are standard. In the US context the TPON network could convey the equivalent of ten × 1.544 Mbit/s streams of traffic rather than the eight × 2.048 Mbit/s assumed in this chapter.

Current work has centred on a transmission system design capable of serving up to 128 network terminations for telephony. However, at the outset, sufficient bandwidth has been made available for up to a total of 120 ISDN or 240 telephony circuits. This corresponds to a system bit rate of 20.48 Mbit/s.

9.3.3 Upgrade scenarios

In simple terms, a TPON network can be upgraded to carry broadband services merely by the addition of an extra wavelength — hence the inclusion in TPON of the optical filter in the network termination to prevent other wavelengths interfering with the TPON receiver. In practice (at least in the medium term), the optical power margins do not allow broadband signals of several 100 Mbit/s to be transmitted through such a high optical splitting ratio as can be used for the lower-bit-rate telephony signals. Broadband signals are therefore injected into the network part-way down the split at the cabinet position utilizing spare ends on the optical couplers as shown in Fig. 9.6; this still allows for an economic level of split to share optical component costs. These issues are discussed in some detail in Chapter 15.

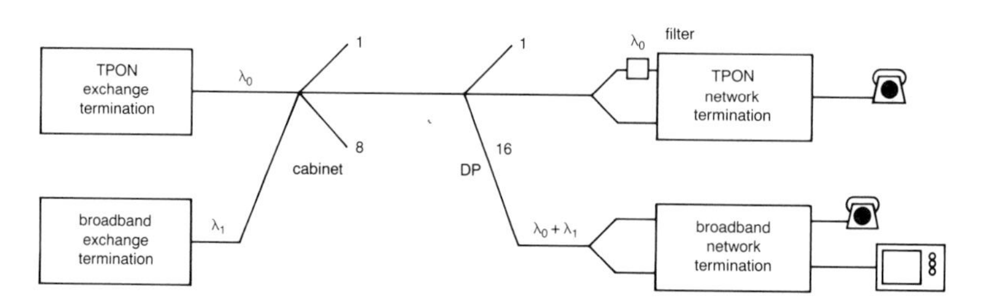

Fig. 9.6 Broadband upgrade.

9.4 COSTS

Initial cost estimates for the mid 1990s for a TPON network with average line lengths are shown in Fig. 9.7. The vertical axis shows the cost per line as a function of the number of telephone lines provided at each network termination. The costs include both the installation as well as the equipment costs, the reference point being a nominal average cost to provide a copper pair. Optical components costs that are significantly cheaper than those currently available have been assumed, but it is expected that the cost targets can be met by novel designs coupled with increasing world market volumes.

The costs in Fig. 9.7 decline with increasing numbers of lines per network termination because the significant fixed costs of the optoelectronics are shared by more telephony lines. Certainly for about 15 lines (a typical minimum size for a street multiplexer), the curve shows that street TPON may be more cost-effective than 15 separate copper connections (in practice there are certain added overheads for housing and powering for a street multiplexer that will erode some but not all of the savings).

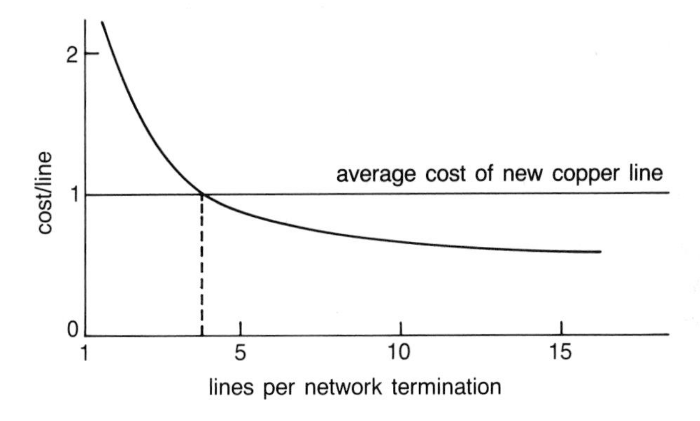

Fig. 9.7 Preliminary cost estimates.

At perhaps five lines per network termination, the costs are still less than for copper, and this suggests that the business TPON market segment could be provided at an early stage. However, Fig. 9.7 indicates that for average line lengths house TPON (i.e. one line per network termination) will not be cheaper than the copper equivalent on the timescale of the estimate, but of course it may be possible to offset some of the added cost by the revenue from future broadband services. As the market volume for optical components increases, so unit costs will fall and the costs of Fig. 9.7 will decrease, but it is too early to say when the costs will fall sufficiently for house TPON to be cheaper than copper. It should not be forgotten that copper network costs are also falling as technology improves installation, maintenance and management. However, on cost grounds there is a clear evolutionary path which suggests that whilst large customer sites currently can be served by point-to-point fibre systems, medium-sized sites (say less than 25 lines) can be cost-effectively served by business TPON in the mid 1990s, with smaller sites being connected as costs fall in the latter half of the 1990s. In parallel, street TPON can begin the process of moving the fibre closer to the residential customer.

Telcos need to cope with growth that is often of an infill nature rather than in large greenfield areas, and the replacement of pockets of the existing copper network that have high maintenance costs. This situation calls for a technology that has a low fixed infrastructure cost, with most costs associated with actual take-up of service. A passive splitter-based network, with its lean use of fibres and flexibility to meet service needs at any outlet without running new cables back to the exchange, fits this need. Figure 9.8 shows that for a four-line business customer, only 25% of the total initial cost is associated with up-front risk money spent on the exchange termination and fibre infrastructure. This emphasizes how both TPON and its upgrade to broadband can meet the needs of a progressively growing network. This should be compared with architectures relying on complex street electronics, where the common electronics and the cabinet must be installed irrespective of how few customers are on the network.

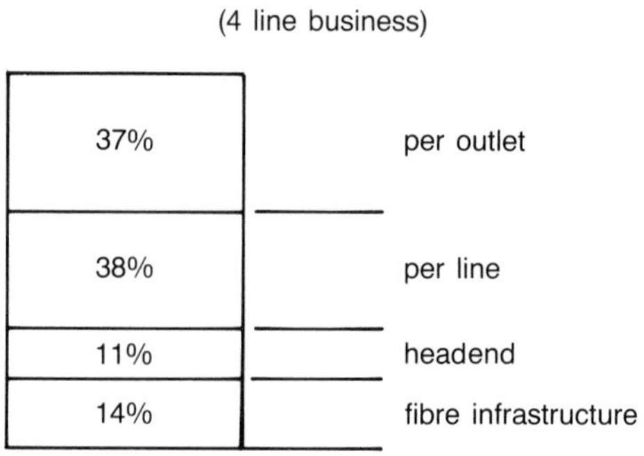

(4 line business)

37%	per outlet
38%	per line
11%	headend
14%	fibre infrastructure

Fig. 9.8 TPON projected cost breakdown (four-line business).

9.5 DESIGN CRITERIA

The following paragraphs describe the design criteria and highlight the key issues involved in the implementation of a TPON system capable of 128 way split. There is a strong interdependency between the electronics design and the optoelectronic/optical plant design. The following description relates predominantly to the electronics and overall system design, whilst the optoelectronic and optical plant issues are covered in Chapter 11.

A system was originally designed to prove the technical feasibility of the TPON approach [5]; more recently a more robust system has been field trialled [6] as part of the Bishop's Stortford fibre trials in the UK, conducted jointly by BT, GPT, BICC and Fulcrum.

9.5.1 Exchange interface

There is a general trend towards statistically multiplexed common-channel signalling, for example in CCITT No 7, I.420 and BT's own specification DASS2. By its nature TPON, like the copper network it replaces, is a logical star and therefore cannot directly use a statistically multiplexed signalling system without significant overheads. One of the main functions of the exchange termination is therefore to convert the exchange-based statistically multiplexed common-channel signalling into a channel-associated signalling system for use over the TPON network. In the demonstration system, a simple bit-assigned system compatible with proprietary PBX internal signalling was used. This is not suitable for practical deployment, as it has no error protection and does not include a comprehensive maintenance capability. The BTS will normally have a very low bit-error ratio, but under fault conditions the error rate may rise dramatically. It is particularly important during the onset of faults that the service to the affected network terminations is shut down in a controlled manner and that any maintenance procedure can be kept in operation as long as possible prior to total failure of the system. The HDLC (high level data link control) layer 2 protocol provides a suitably robust framing and error-protection system and is readily available. HDLC also allows a flexible message set to be implemented allowing as yet unspecified future enhancements.

In future, TPON systems will need to interface with the new standards which are emerging such as Q.931 and SDH (synchronous digital hierarchy). Q.931 is easily accommodated since it is based around 2 Mbit/s tributaries broadly similar to DASS2, requiring only a different common-channel to channel-associated signalling conversion at the exchange termination. Although there are many parts of the SDH specification which are not yet complete, it is clear that the TPON concept is readily compatible in that, for example, the traffic from up to seven 20.48 Mbit/s TPON systems can easily by multiplexed together to interface to the exchange over a 155 Mbit/s SDH link.

9.5.2 Optoelectronics and fibre network

The optoelectronics and optical plant for a passive optical network are described in some detail in Chapter 11 but the following is a summary of those aspects which are essential to a full appreciation of the electronics design of TPON.

The TPON optical window required is determined by the production and operational tolerances of the filter in the network termination, the wide temperature range experienced in the customer's premises and the spectral characteristics of the exchange laser. This window should be as narrow as possible in order to reserve spectrum for future system evolution. Thus, in addition to a high mean output power (about 1 mW), the exchange transmitter is required to be single longitudinal mode with tight control of the centre wavelength (<1 nm). This implies distributed feedback (DFB) laser technology with temperature control and high laser-to-fibre coupling efficiencies.

The exchange receiver needs to cope with varying bit-to-bit pulse amplitudes which arise because of different path losses and imperfect correction by the automatic levelling system in the BTS. For this reason, d.c. restoration circuitry is used. This copes with the varying d.c. content of the customer-to-exchange bit stream produced by the varying number of customers simultaneously using the system.

Spectrum conservation in the customer-exchange direction is sacrificed for economy, since in the medium term most of the broadband services are unidirectional. A relatively simple design of network termination transmitter can be used based upon a Fabry Perot cavity, ridge or buried heterostructure laser device. An important objective of the design of the BTS frame structure is to minimize the on-time of the laser, reducing thermal stress and avoiding the need for a thermoelectric cooler which is both expensive, power-hungry and a reliability risk. Indeed, some thought has been given to coding schemes for speech which minimize the number of pulses sent, but the relative advantages were found to be too small and unpredictable to warrant the extra complication. A further simplifying feature is that the back-facet monitoring diode is not required because power monitoring and control are carried out by the exchange termination via the BTS.

Wavelength-flattened optical splitters are required in order to achieve transparency in both the 1300 and 1500 nm windows. Also, to minimize insertion-loss difference between the ports of the couplers, it is necessary to have good control over coupling ratio tolerances of the elemental 2×2 couplers.

9.5.3 BTS hardware

The BTS (and associated optical components) provide a transparent path for the traffic data carried over the network. This traffic data consists of:

- service carried (e.g. 64 kbit/s speech);
- signalling to control the service (e.g. loop disconnect and ringing carried as 'messages');

- maintenance signals for the service (e.g. activate loop).

The BTS frame structure has been designed to carry traffic data arriving at eight 2.352 Mbit/s PCM ports at the exchange; in addition capacity needs to be allocated for BTS control and ranging. For simplicity, the frame structures in both directions have similar formats but differ in functional detail. Also data in the broadcast direction is scrambled to facilitate clock recovery at the remote terminations. There is no need for the data from the network termination to be scrambled because the system is totally synchronous and clocks are available at the BTS-Master.

A block diagram of the BTS-Master is shown in Fig. 9.9; key elements are the formatter, ranging computer and housekeeping computer, the functions of which are covered in more detail below.

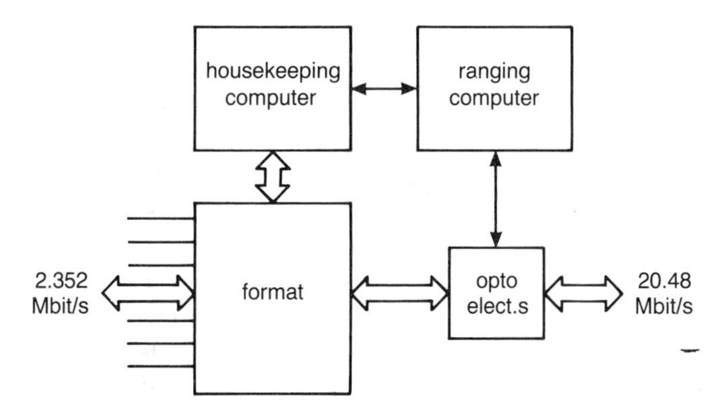

Fig. 9.9 BTS-master block diagram.

A block diagram of the BTS-Slave is given in Fig. 9.10. The timing circuit recovers timing and provides the programmable delay function. A 'circuit engine' provides a programmable multiplexer/demultiplexer to deliver the 8 kbit/s circuits required by the network termination — for example for telephony the circuit engine can be programmed to extract nine particular 8 kbit/s circuits from the multiplex. For a multiline network termination only one timing circuit is required together with a circuit engine per line. Each circuit engine must access contiguous channels in the original 2.352 Mbit/s multiplex at the exchange end, but different circuit engines (even connected to the same timing circuit) may access different 2.352 Mbit/s streams which may be carrying other services. Thus grooming can be carried out at the network termination. In the current design the number of 8 kbit/s circuits available from a circuit engine is limited to 20 (corresponding to 160 kbit/s — perhaps an ISDN service). Circuit engines could be designed with larger

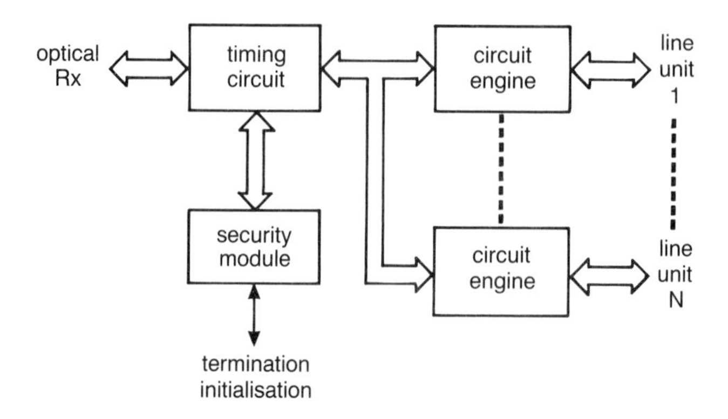

Fig. 9.10 BTS-Slave block diagram.

capacities and more complex grooming capabilities, for example to enable a 2 Mbit/s PBX to receive several telephony channels and a few private circuits, any unused channels in the 2 Mbit/s multiplex being left unassigned.

9.5.4 TDMA frame structure

The system bit rate is 20.48 Mbit/s and the frame repetition rate is 100 Hz. 20.48 Mbit/s was chosen as the nearest integer multiple ($\times 10$) of the European network frequency standard 2.048 Mbit/s; 100 Hz (10 ms) framing was selected as a compromise to minimize control response delays whilst not introducing too much framing overhead for the control, ranging and synchronization functions. The transmission loop delay of the system is about 1 ms for any fibre length up to the ranging limit of 20 km.

The frame structure in the broadcast direction is shown in Fig. 9.11. The synchronization frame consists of 5120 bits, of which 196 sequential nulls (logic '0' prior to scrambling) represent a distinct synchronizing pattern which defines the reference point for the multiframe; the remaining 4924 bits in the synchronization frame have not been allocated but could be used for optical plant testing and maintenance. Traffic and control channels are broadcast in the remaining 199 680 bits which have been partitioned into 80 'basic frames', each consisting of 2352 traffic data bits and 144 BTS control bits.

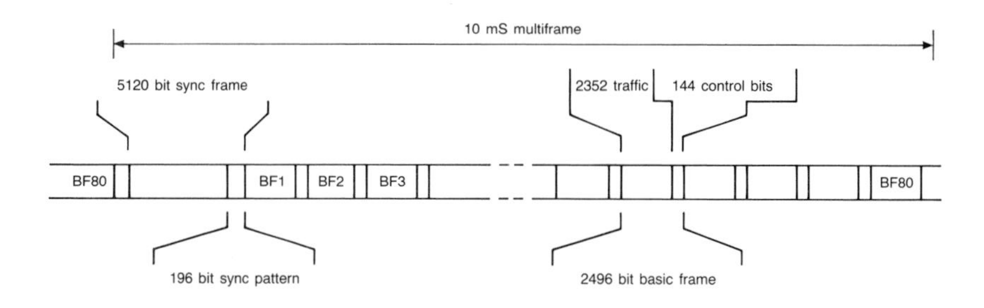

Fig. 9.11 Broadcast frame structure.

In the return direction (Fig. 9.12), bits arriving at the exchange are correctly interleaved and precisely positioned in the multiplex by a continuous ranging function. The period corresponding to the broadcast synchronization frame is reserved in the return direction for the ranging function; the first 4096 bit periods have been reserved for the 'phase 1' ranging pulse, 304 bits have been reserved for DC restoration within the exchange optical receiver, and 720 bits have been reserved for 'phase 2' ranging pulses. The remaining 199 680 bits are partitioned for traffic data and BTS control in the same manner as for the broadcast multiplex.

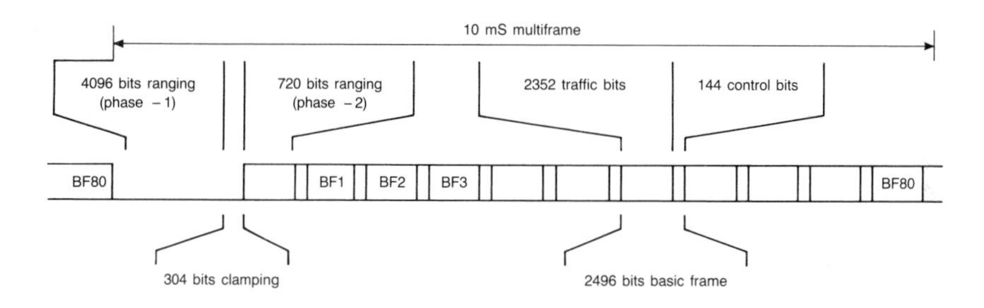

Fig. 9.12 Return frame structure.

9.5.5 BTS ranging

Ranging operates in two distinct modes — 'phase-1' ranging which occurs only during installation of the termination (or after fibre maintenance) to determine the fibre loop delay accurate to within ± 1 bit (50 ns) and 'phase-2' ranging which operates continuously, achieving an accuracy of 1/10th bit (±5 ns). During normal operation phase-2 ranging continuously tracks any minor changes in phase or amplitude.

Each new network termination is progressively brought into service from the BTS-Master in a controlled manner. The procedure is as follows.

- Network termination is physically connected and labelled as network termination number 'N' awaiting service (N in range 0 to 127).
- The BTS-Master requests network termination N (when it is available) to commence phase-1 ranging via the housekeeping channel.
- Once phase-1 ranging is completed satisfactorily the BTS-Master can request phase-2 ranging to commence.
- Once phase-2 ranging is completed satisfactorily the BTS-Master can instruct traffic to be carried.
- A security interchange then takes place between the network termination and the exchange termination using a procedure akin to a PIN (personal identification number) to ensure that the installation is authorized.

9.5.5.1 Phase-1 ranging

The termination to be ranged responds to a command from the BTS-Master to transmit a single pulse with maximum electronic delay relative to the received broadcast multiplex (250 μs). The pulse will therefore arrive at the exchange within the guard period allocated for phase-1 ranging — delayed by an amount which corresponds to the loop delay for that particular termination. The exchange measures the delay and sends a command to the termination to insert the compensatory delay value required to fix the total loop delay to about 250 μs. For a 20 km reach the compensating delay would be near zero, while for a network termination very close to the exchange the compensating delay would be near 200 μs.

Phase-1 ranging determines the distance to within 50 ns (10 m of fibre) and the phase-2 ranging process calibrates the delay to within 5 ns (1 m of fibre).

9.5.5.2 Phase-2 ranging

Dedicated timing subslots have been allocated for phase-2 pulses for each network termination because it is necessary to calibrate the loop delay continuously for each network termination to compensate for temperature variations and component ageing. Each timeslot is five clock pulses wide (250 μs) to accommodate errors in phase-1 ranging. The phase-2 ranging process determines the loop delay to within 5 ns and a command is sent to the slave to insert a compensatory delay to bring the overall loop delay to precisely 250 μs.

During phase-2 ranging the amplitude of the received pulse is also measured and similarly remotely adjusted via the housekeeping control channel to ensure that a consistent level is received from all network terminations. Phase-2 ranging must run continuously since the system delay is subject to variation mainly due to temperature — for example, fibre propagation delays vary up to 0.22 ns/km/°C.

In order to carry traffic the BTS adds its ranging delays to a delay equal to the effective bit position in the multiplex to ensure that precise bit interleaving takes place.

9.5.6 BTS housekeeping

The above description of the BTS included reference to a housekeeping or control channel. The system has 144 separate 8 kbit/s control channels, one for each possible optical network termination plus additional maintenance channels. Each 8 kbit/s control channel is formatted into a multiframe lasting 10 ms and containing 80 bits as shown in Fig. 9.13, each bit being carried in each of the 80 basic BTS frames in the BTS multiframe of Fig. 9.13. It was decided at an early stage of the development that the transfer of configuration and control data across the network should be in a bit-mapped format rather than as a message-based protocol in order to reduce significantly the complexity and cost of the network termination.

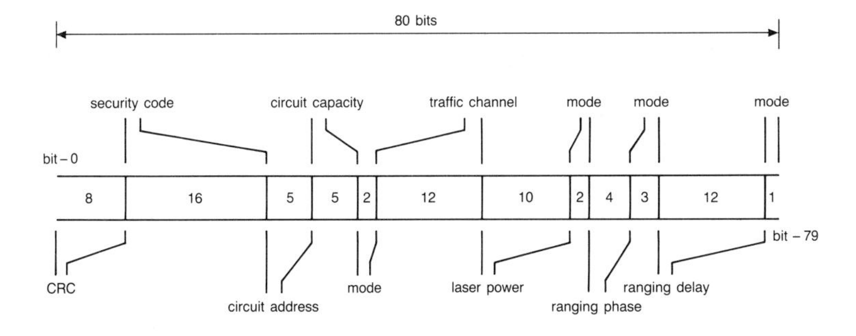

Fig. 9.13 Housekeeping multiframe.

9.5.6.1 Error protection in the control channel

It is a requirement that the TPON system operates down to a 1 in 10^3 error rate. Telephony in the presence of 1 in 10^3 errors will be intelligible, but it is likely to be the performance of BTS housekeeping under such conditions which will govern the way in which the system degrades. Housekeeping data in transit across the network is therefore protected by an 8-bit CRC (cyclic redundancy check) transmitted with the control data to each BTS slave. The CRC is derived from the polynomial $(X^8 + 1)$ which detects 100% of error burst length eight or less, 99.2% of error burst nine and 99.6% of error burst ten or more. As additional protection, a persistence check is performed for received-error-free control words, the result of which can optionally be used to disable the laser transmitter under excessive error conditions when there is a high probability that undetected control errors will occur.

Future versions of the TPON BTS may justify a more complex error protection strategy (with increased system overheads) should this be shown to be required after operational performance analysis.

9.5.6.2 Housekeeping control systems

The BTS housekeeping control systems have been designed to service up to 128 network terminations with each termination allocated a bi-directional 8 kbit/s channel. There are 16 additional 8 kbit/s channels which are currently unallocated but which may be used for maintenance purposes. The overall data transfer rate required to support 128 + 16 housekeeping channels is 2.304 Mbit/s and the instantaneous housekeeping bit rate equals the system bit rate (20.48 Mbit/s) necessitating hardware buffers to couple the control systems synchronously to the TDMA multiplex. In practice, the rate of change of housekeeping data will be quite low and conventional microprocessor architectures can be applied.

The housekeeping control system communicates with the network management hierarchy via a single addressable serial interface and provides the software functions necessary to keep the network operating, detect faults and allocate service to new customers. These functions will normally be accessed via the network management hierarchy but for testing/commissioning, the BTS Master has a terminal interface and simple command handler to enable a subset of the housekeeping functions to be invoked manually.

9.5.7 Choice of bit interleaving

In the above sections it was assumed throughout that the signals from different network terminations are bit-interleaved; an alternative would be to group bits and interleave bytes or even larger packets of bits. The disadvantage of bit-interleaving is the need to have tight control of the ranging process and careful design of the exchange and receiver. The primary advantages of bit-interleaving are as follows.

- The low mark space ratio of the network termination laser removes the need for a peltier-cooler to prevent the laser temperature rising during sequential pulses. This results in lower power, lower cost and more reliable network termination.
- It easily provides the necessary 8 kbit/s granularity.
- It reduces the complexity of the multiplexing functions in the BTS.

Overall, the balance is in favour of bit-interleaving mainly because of the cost savings, the problems of ranging and receiver design having been demonstrably overcome by the building of trial systems.

9.5.8 Network diagnostics features

To facilitate network fault-finding during initial trials of the TPON system when the frequency and type of faults would be largely unknown, it was thought expedient to include relatively powerful diagnostics capabilities within the BTS. An analogue sampling subsystem was therefore included as an integral part of the ranging system within the BTS Master.

This sampling subsystem utilizes a special analogue port at the exchange optical receiver which in effect provides a linear window on the upstream TDMA multiplex prior to data slicing. This port is sampled by a high-speed analogue-to-digital converter at a range of 81.92 MHz, phase-locked to the system clock. The embedded controllers within the BTS use this sample data to implement the remote laser-levelling servo loop. In addition, optical network parameters can be measured to detect, for example, reflections which might occur under fault conditions in a duplex-configured fibre network. The data is made available on demand for processing by network management.

A diagnostics facility implemented for the Bishop's Stortford trial has been a digital storage oscilloscope display as shown in Fig. 9.14. The top trace shows the sliced data presented to the BTS by the exchange receiver, the bottom trace the sampled analogue data including a locally injected calibration 'marker' pulse (approximately 150 ns wide).

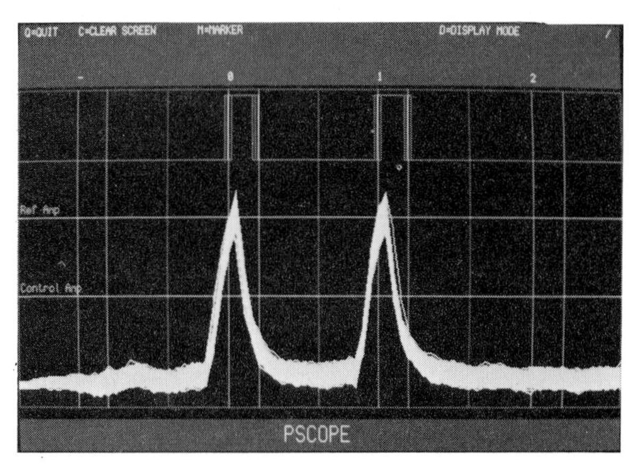

Fig. 9.14 BTS diagnostics trace showing (top) sliced data from exchange optical receiver and (bottom) sampled analogue data (150 ns calibration pulse at left).

9.6 NETWORK TERMINATION

In simple terms the network termination allows an ordinary telephone to be interfaced to the optical network. In practice it performs a range of functions:

- analogue telephony interface (including up to 40 mA line feed and 75 V a.c. ringing);
- optical interface (20.48 Mbit/s);
- converts analogue telephone interface to 64 kbit/s plus signalling;
- controls the laser power within 0.5 dB;
- controls laser pulse position ±5 ns;
- multiplexes/demultiplexes $n \times 8$ kbit/s traffic channels from the 20 Mbit/s;
- operates a defined initialisation and security procedure;
- provides battery back-up in the event of mains fail;

- includes self-test features.

The design of the network termination is a major challenge since the above functions have to be provided at minimum cost (the cost cannot be shared with other customers) and minimum size (it must be unobtrusive if it is to be acceptable to customers). To shorten development time, the network termination used in the Bishop's Stortford field trial was based upon existing multiplexer hardware and was fairly large as shown in Fig. 9.15. This was acceptable since the main aim of the trial was to assess technical features and not market acceptability. Current studies indicate that in the longer term the network termination could be reduced in size to a unit about 100 × 100 × 50 mm plus a separate power supply and battery housed in a plug-top unit similar to that used for calculators. A mock-up of such a unit is shown in Fig. 9.16. This size reduction is achieved by the use of custom ICs, sophisticated interconnection techniques and by careful attention to the optical components.

Fig. 9.15 House TPON network termination for Bishop's Stortford trial.

A key contributor to the size of the equipment is the need for the telephony interface to mimic all the features normally found via a direct copper exchange line, in particular:

- 50 V idle line feed, up to 40 mA feed current;
- 70 V, 25 Hz unbalanced ringing;
- optional attachment of 50 Hz subscribers private meters (SPM);
- 20 dBm howler;

- test access;
- operation for perhaps 1 hour after 10 hours mains fail;
- safe in the presence of mains voltages and lighting surges on the telephony pairs.

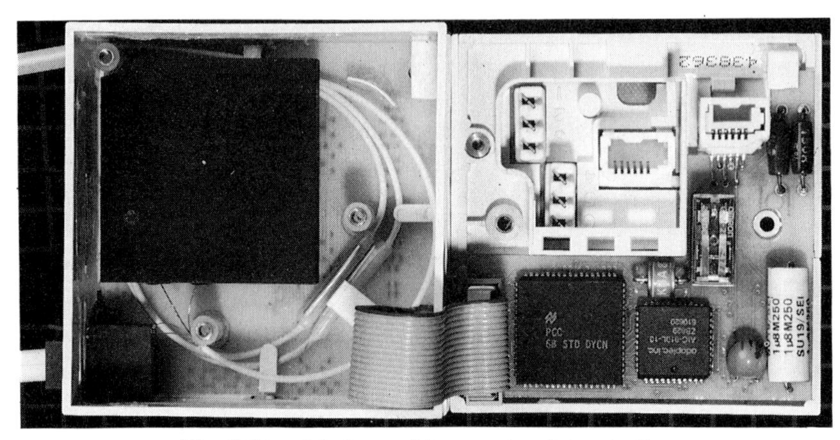

Fig. 9.16 Mock-up of future network termination.

Most of the above items have a long history and are a statutory requirement or are implied in BSI (BS6305) and soon in EEC NETs (Norme European de Telecommunication). However, they represent neither a clear network design specification nor an optimum technical solution and, although not inconvenient for a copper network, can seriously mar the viability of any fibre-based network. Despite this the current design work for TPON is targeted on copying the standard copper interface. Indeed the specification of an interface to provide a telephony service via a fibre network is long overdue for international study and agreement. Currently, a choice exists between existing analogue specifications which are different in every country, ISDN I.420, and perhaps a forthcoming interactive home systems specification (based upon communication between video recorders and hi-fi systems but including a speech capability). None of the above has been tailored to the needs of cost-effective telephony distribution via fibre.

9.7 NETWORK MANAGEMENT AND MAINTENANCE

Any development of the local network should pay particular attention to management and maintenance. Although this added functionality increases the initial capital cost it can reduce the whole life costs of the system, especially if the system approaches the ideal 'hands-off' network which requires no manual intervention. The current copper network is far from being a hands-off network since manual access is involved at the exchange (on the main distribution frame (MDF)), at cabinets and at pillars, as well as at the customer's premises to provide new circuits. Each time the MDF, cabinet or pillar is disturbed there is the possibility of disturbance to other customers' connections. Although the copper network can be modified to overcome some of these deficiencies, a broadband passive optical network inherently overcomes them since new circuits can easily be provided just by adding extra equipment at the network termination. The following paragraphs highlight the key issues which are directly pertinent to the TPON architecture and to the associated TPON element manager software.

TPON element manager software will typically have the following functionality:

- maintaining a complete record of all the TPON elements under its control (e.g. circuit card details) — such a set of records might cover all the systems within an exchange serving area and might therefore include more than a hundred TPON optical networks;
- managing remote access to the TPON elements for configuration, testing, traffic control, analysis and maintenance activities;
- monitoring the TPON elements, performing limited processing of events, performance and usage information, passing selected events to the network control level;
- providing concentration facilities for management messages.

The TPON element manager must be able to communicate with the network level controller. This can either be by direct connection or via transparent messages in the DASS2 connection to the local exchange.

There is much about the TPON architecture that simplifies maintenance.

- The network termination is in constant contact with the exchange termination and can report local faults, e.g. self test, line card faults, low battery charge.
- The exchange termination is continuously monitoring the laser power and delay for each network termination — hence progressive degradation of lasers and the fibre network can be spotted early and ideally action taken before the fault affects service.

- The signalling capability between the exchange and the network termination enables extra maintenance features such as loop back to be added easily — this signalling capability can be used to form a direct link between the network, its configuration and the field staff to avoid errors in network records and transferring works instructions.
- Correlation of faults from several network terminations can help to pinpoint faults in the optical network — for example if all the terminations from a particular DP have failed it is likely that the fault lies between the cabinet and the DP.

These features together with the optical plant maintenance aids described in Chapter 11 ensure that a TPON network has a good degree of maintainability. However, many of the proposals need to be verified in the field before the TPON approach can claim to have solved all the problems. The Bishop's Stortford field trial will provide valuable data in this area [8].

In the longer term it may be possible to utilize the BTS channels to provide additional communications channels for the maintenance engineer via 'clip-on' technology to gain access to a low-bit-rate data channel and communicate directly with the network-management computers.

9.8 RELIABILITY AND SECURITY

9.8.1 Reliability

A key parameter for any network is its reliability. The current targets for TPON are perceived to be 10-year mean-time-to-failure (MTTF) for the link between a single customer and the exchange and a 60-year MTTF for faults affecting all 128 customers. Detailed calculations have shown that these targets can be met if the following additional design features are included:

- duplication of the exchange termination laser;
- network terminations in rural areas having a 24 hour rather than 10 hour stand-by battery;
- overhead drops avoided in favour of underground feeds.

The exchange laser can be duplicated by taking a second fibre from the stand-by laser to the spare coupler ends at the cabinet position. This can also be used to provide some degree of diverse routeing to overcome the effects of mechanical damage to one of the cables. The need for a 24 hour battery will add to the size of the network termination but it is only needed in rural areas where the mains supply is less reliable. It may be appropriate to allow customers to rent as much battery capacity as they feel they personally desire; this allows the customer to make the compromise between cost, size and reliability. Overhead drops are a problem for copper as well as fibre and underground connection is always preferred.

9.8.2 Security

Initially it was feared that the TPON network might not be sufficiently secure from eavesdropping or malicious interference because it broadcasts traffic data to all customers. However, detailed studies have shown that the network is in fact more secure than the current copper network because:

- the network termination can only be connected by authorized staff with personal PINs using equipment with built-in identification codes which are checked by the exchange termination;
- the network is continuously monitored for optical power level and delay, and interference with the fibre will be instantly detected;
- the optoelectronics and BTS at the network termination are in a shielded unit which only allows access to timeslots under control from the BTS Master — customers are therefore denied direct access to both the optical broadcast signal and the electrical multiplex;
- there is no user-accessible optical connector at the customer's premises.

Clearly any customers requiring extreme levels of security can provide additional encryption for their data as in the current network.

9.9 CONCLUSION

TPON represents an exciting early opportunity to deploy an all-fibre network economic for multiline telephony that also resolves many of the operational problems of the copper network. When these benefits are coupled with the ease of provision of future broadband services, the economics may fully justify the provision of fibre to single-line residential users. However, this promise is not without its risks, and the uncertainties in costs, reliability and maintainability need to be more fully studied before radical decisions are taken. It is for this reason that the Bishop's Stortford UK field trial is in hand. The UK is not alone in wanting to know the answers to these issues, and following the success of the UK trial, other similar trials are now operational or planned around the world, for example in the USA, Japan, Germany and the Netherlands, involving all the leading world manufacturers. Meanwhile BT intends to build on its lead and prepare for the next generation of equipment, mainly aimed at fibre to the small and medium-sized business.

REFERENCES

1. Ritchie W K: 'The local distribution network — an overview', BT Technol J, 7, No 2, pp 7-16 (April 1989).
2. Dufour I G: 'Flexible access systems', ISSLS'88, Boston, USA (1988).
3. Foxton M G and Pilling G A: 'Optically coupled remote multiplexers', BT Technol J, 7, No 2, pp 55-64 (April 1989).
4. Hoppitt C E: 'From copper to fibre in easy stages', IEEE Communications Magazine, 24 , No 11 (November 1986).
5. Oakley K A, Taylor C G and Stern J R: 'Passive fibre local loop for telephony with broadband upgrade', ISSLS '88, Boston, USA, pp 179-183 (1988).
6. Hoppitt C E, Astbury M L, Keeble P and Chapman P A: 'Operations and maintenance experience of the Bishop's Stortford fibre trial', 3rd IEE conference on Telecommunications, Edinburgh, pp 165-169 (March 1991).
7. Hoppitt C E and Rawson J W D: 'The UK trial of fibre in the loop', British Telecommunications Engineering, 10 , Pt 1, pp 45-58 (April 1991).
8. Hoppitt C E: 'The application of passive optical loops', Forum '91, Geneva, pp 21-24 (October 1991).

10

BROADBAND SYSTEMS ON PASSIVE OPTICAL NETWORKS

D W Faulkner and D I Fordham

10.1 INTRODUCTION

For regulatory and technical reasons, cable TV and telephony systems have evolved separately. More recently, the requirement for interactive TV and other services has led to the design of local access systems with a large downstream video capacity and sufficient upstream capacity to carry the data needed for programme selection, interactive data services, telephony, and possibly an upstream video channel. Until now, telephone and cable TV companies have not been well placed to provide mixed service capability on the same network, for both regulatory and technical reasons. With the passive optical network (PON), however, a medium with potentially enormous capacity is made available which can be largely justified on the basis of telephony service provision, but which can be readily upgraded, through the application of wavelength division multiplexing (WDM), to provide a range of broadband services merely by adding (or changing) terminal equipment.

In Chapter 9, various system options for providing telephony over a passive optical network (TPON) were described. Figure 10.1 illustrates, in principle, how such a network can be upgraded to provide downstream broadband services using an additional optical wavelength (λ_1). Initially the use of WDM would be limited to the addition of one or two wavelengths but, as the technology matures and the need arises for increased capacity to carry new services such as high-definition television (HDTV), this could be extended in an evolutionary way by the application of more wavelengths

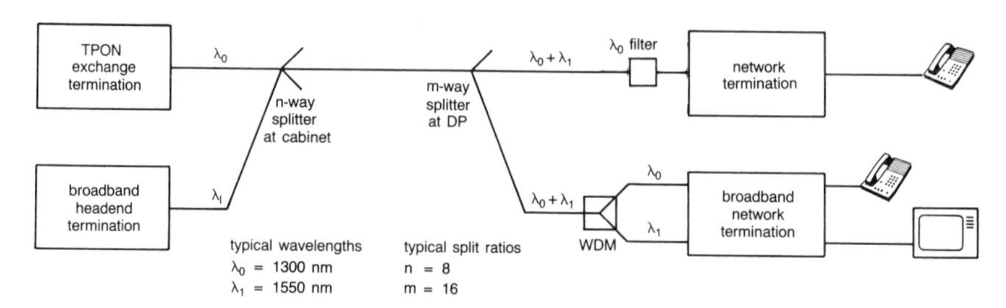

Fig. 10.1 An upgrade from TPON to BPON.

and a more advanced WDM device (see Chapter 15). Thus, by careful design, it should be possible for a single, unchanging, network to satisfy economically the requirements for telephony and broadband services both in the short term and well into the future.

This chapter considers the customer requirements for the provision of broadband services on a PON and discusses methods of practical implementation of the terminal equipment necessary for a broadband PON (BPON). The prime application is assumed to be as a cable TV upgrade to TPON, although the distributive nature of the network is also suited to the provision of cable TV services alone.

10.2 NETWORK OUTLINE

The detailed design of a PON and descriptions of the necessary components are covered in Chapter 11, but a brief description of the network as it relates to a broadband upgrade is useful here. So that minimum costs can be achieved in the short term for telephony, it has been assumed that TPON would occupy the whole of the 1300 nm window (1260-1340 nm), reserving the 1550 nm window for additional use. With existing technology it is possible to use four wavelengths in this window, one for maintenance access and three for broadband upgrades. Figure 10.1 shows the addition of a single 'broadband' wavelength and it can be seen that it is added at a point in the network which introduces a lower optical-splitting ratio. This is necessary since the greater bandwidth required by the broadband services results in a lower optical budget being available. In practice this different splitting ratio can be accomplished easily because the initial stages of splitting can be made up of elemental 2*2 optical couplers which provide additional input ports at various levels of split.

For the broadband upgrade to take place non-intrusively, a bandpass filter must be present at each TPON terminal from the outset. This eliminates the crosstalk problems which would otherwise arise at the TPON receiver when the extra wavelengths are added. For those customers requiring broadband services, the optical spectrum is first divided into the 1300 nm and 1550 nm windows using a coarse-grained WDM device as Fig. 10.1 shows. The presence of further 'broadband' (or maintenance) wavelengths will require additional optical-bandpass filtering.

10.3 SYSTEM REQUIREMENTS

10.3.1 Services

The primary role of cable TV networks is to provide a wide range of TV channels to domestic customers. Programme material is at the moment limited to NTSC, PAL and SECAM, but this situation will change considerably over the next few years. Stereo sound is becoming available, and less compatible changes are imminent. Multiplexed analogue component (MAC) channels, offering better picture quality and improved transmission performance, are available on some satellite transponders, and HDTV systems are under investigation. Cable systems will be required to carry all these types of channel as they become available although full capability may not be needed at the outset.

In addition to the one-way entertainment services described above, it is likely that networks will be asked to carry an increasingly wide range of services, for example:

- telephony and data;
- upstream signalling to allow interactivity;
- on-demand TV programmes (video library);
- information services including picture videotex;
- upstream contribution video, videophony or surveillance.

Although aimed to some extent towards domestic customers, some of these services can be expected to appeal more directly to the business community where the cost constraints, so crucial for entertainment TV, are not so demanding.

On any network it is essential to protect the cable operator's revenue by applying conditional access to some services whilst others are available with no restriction. This conditional access can be achieved either by the use of scrambling or switching (or both) and would depend to a large extent on the type of network structure. Clearly the purely distributive tree-and-branch coaxial networks rely on scrambling whereas switched-star structures can, through software control, provide inherent conditional access.

10.3.2 Performance

National regulatory authorities specify the minimum acceptable performance criteria for cable TV networks. However, these requirements are aimed particularly at coaxial tree-and-branch and do not deal with some of the parameters which are found to be important in newer systems, particularly where modulation or demodulation occur within the network. The parameter which is usually regarded as providing the best measure of network performance is the signal-to-noise ratio (SNR), often quoted as carrier-to-noise ratio (CNR). The relationship between them is dependent on the type of modulation used. It is important to remember, however, that SNR is not the only impairment and other distortions such as crosstalk and sinewave interference can be far more disturbing and therefore subjectively more significant. The CNR performance limit for VSB-AM transmission of PAL is 43 dB which translates to a luminance-weighted SNR also of 43 dB. In coaxial tree-and-branch networks, with their cascades of line amplifiers and splitters, it is SNR which is particularly critical and networks are designed around it. For passive optical networks, where line-transmission equipment would only be present at the terminal ends, a given SNR should be easier to achieve and maintain. Furthermore, the opportunity exists to give better performance than is currently expected of cable TV networks. It is proposed therefore that a target figure of 48 dB luminance SNR be used for PONs. This figure includes all contributing elements in the network — headend (including off-air demodulators), transmission system, and in-house equipment and distribution.

The use of SNR as a measure of performance may in practice be limited to analogue systems. If digital transmission were to be used, SNR (as well as all other traditionally quoted impairments) would be determined purely by the encoding and decoding process. Transmission performance would, as with other digital systems, be characterized by bit error rate (BER) which is subjectively very different in its effect to SNR. A BER of 1 in 10^9 is often used as a target in characterizing digital systems even though a much worse figure (1 in 10^6 perhaps) would be acceptable for video transmission.

10.4 FIBRE IN EXISTING BROADBAND NETWORKS

Existing cable TV distribution systems are based upon one of two topologies — switched-star or tree-and-branch. In the switched-star network, the primary distribution network terminates near a group of customers at a remote switch point. This part of the network can use either coaxial or optical transmission, the latter having been chosen in the BT network used in the Westminster franchise area [1]. At the remote switch, TV channels are selected by the customer and routed over a dedicated 'secondary' link. Currently these links use coaxial cable, since the cost of the optoelectronics required for a fibre to each home is prohibitive. However, the challenge of reducing these costs to a level comparable with coaxial technology is being addressed which may result in the emergence of a cost-effective all-optical switched-star network (see Chapter 8).

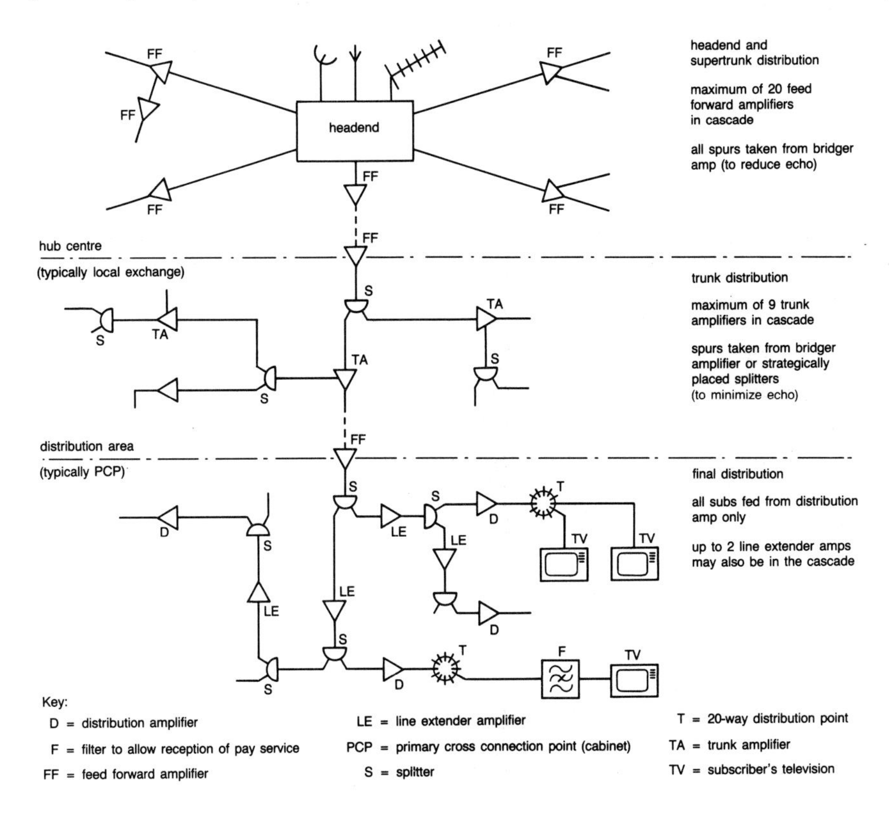

Fig. 10.2 Basic 'tree-and-branch' cable TV network.

Tree-and-branch networks, as shown in Fig. 10.2, operate in broadcast mode where a single headend transmitter is used to feed a multiplex of all channels to all customers. The signal is amplified at intervals to overcome cable and splitter losses. These chains of amplifiers and splitters not only degrade the signal but also represent a very significant reliability hazard. It is for these reasons that there is great interest, particularly in the USA, in upgrading at least the primary distribution network with low-loss optical fibre, thus eliminating large numbers of amplifiers with their need for remote power feeding.

Such upgrades would be most attractive if direct connection, without costly remodulation equipment, could be made between fibre and coaxial systems. VSB-AM modulation found on coaxial networks is not ideally suited for use on optical fibre owing to SNR limitations and intermodulation distortion imposed at the optical transmitter. Lasers are being developed, however, which offer the prospect of meeting the required performance in these two areas, and, although the optical power budget achievable to date has been limited, in-line optical amplifiers have the potential for improving this to the point that optical splitting of the signal is viable. Provided that adequate and sustainable performance can be achieved from these devices then the benefit to tree-and-branch networks will be enormous.

10.5 BROADBAND PASSIVE OPTICAL NETWORKS

10.5.1 Equipment constraints

Functionally a BPON is similar to a coaxial tree-and-branch network in that a large number of TV channels are transmitted to a large number of customers. These are conflicting requirements since increasing the number of channels (and hence bandwidth) reduces the optical power budget available with given optoelectronic components, which in turn reduces the degree of optical splitting that can be achieved (and hence the number of customers served). This conflict is most significant for the optical receiver required for each customer, since receiver sensitivity is inversely proportional to the bandwidth. The receiver design is therefore critical because it very significantly affects both network performance and cost. The design of channel demultiplexer and signal demodulator (if required) in the customer's premises is also important because their costs cannot be shared. Equipment at the headend is not as cost-sensitive since it is shared amongst a number (dependent on splitting ratio) of customers.

Three modulation techniques are considered for the distribution of cable TV services over optical fibre — AM subcarrier, FM subcarrier and digital TDM.

10.5.2 Amplitude modulation

As already described, the use of AM modulation on fibre leads to a limited optical power budget, and PONs of small split size. However, the emergence of fibre amplifiers, to some extent, overcomes this difficulty by allowing a low-level signal to be amplified at suitable points along the route. The optical splitting ratio can then be increased, provided that the linearity and SNR through the system are adequate.

Since the network now contains active devices, it is known as a transparent optical network (TON), as opposed to a PON which contains only passive components. The design philosophy of TONs differs from PONs in a number of ways which could alter the future evolution towards fibre in the local loop. From the perspective of a cable TV provider, the use of a TON in the primary distribution network could prove economical soon and the extension of fibre direct to the customer could follow with the addition of low-cost optical amplifiers in the secondary distribution network. A telephone company, on the other hand, may opt for the relative simplicity of a PON and provide broadband upgrades via additional wavelengths, provided sufficient power budget is available. The broadband power budget problem can be significantly eased by considering alternative modulation schemes such as FM or digital.

10.5.3 Frequency modulation

Frequency modulation gives the well-known advantage of requiring lower CNR than AM for a given SNR. For example, using frequency deviation of 16 MHz/V gives an FM advantage of about 32 dB (electrical) for PAL signals, which translates to an optical advantage of 16 dB. With this improvement, optical splitting becomes practical. Of course the penalty to pay for using FM is bandwidth increase from about 6 MHz for AM to 30 MHz for FM, but since bandwidth is relatively cheap with currently available single-mode fibre systems, this does not represent a serious problem.

Frequency modulation also offers a further advantage when numbers (say 16 and above) of channels are multiplexed together. The unrelated modulated signals add together to form a composite waveform which has relatively few large voltage excursions. These can be removed (clipped) prior to driving the laser over the full extent of its linear operating characteristic, thus increasing the transmitted SNR. Although the clipping generates distortion products, these are at a low level and, provided that the modulated carrier frequencies are contained within one octave of the frequency spectrum, the second-order products fall out of band and do not affect picture quality. In fact it has been shown that 400% overmodulation (signal peaks four times larger than needed to utilize the laser's linear operating region fully) is possible and brings about a 6 dB improvement in optical budget. To carry the required number of TV channels and accommodate them within an octave of the frequency spectrum necessitates the use of carrier frequencies in the gigahertz region. However, high-speed lasers and receivers are available which allow transmission on fibre at these rates. The technique of transmitting a multiplex of carriers (actually subcarriers) on an optical wavelength (the carrier) has become known as subcarrier multiplexing (SCM) and is stimulating interest worldwide.

Figure 10.3 shows a possible network configuration for an analogue (SCM) BPON. A multiplex of 32 video-modulated carriers is assembled at the headend, occupying the frequency band between 950 MHz and 1.75 GHz. To achieve this packing density, a channel separation of 24 MHz and an FM deviation of 13 MHz/V are required. The deviation could be increased (to say 16 MHz/V) at the expense of increased channel separation and thus fewer channels. The customers' equipment consists of an optical receiver followed by a standard satellite set-top receiver unit to select and demodulate the required channel.

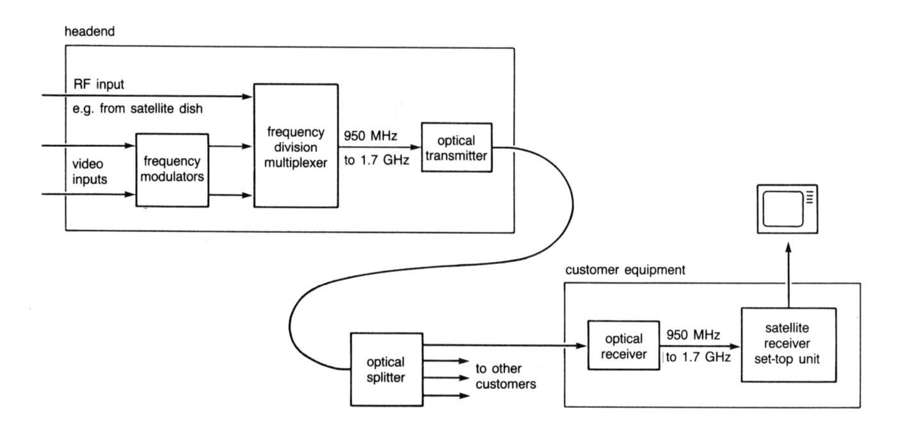

Fig. 10.3 Network configuration for SCM BPON.

The optical power budget and degree of passive splitting possible are dependent on laser launch power and receiver sensitivity. In general, as these parameters increase, so do terminal equipment costs. However, this need not lead to higher costs, since the improved power budget made available can be used to increase the splitting ratio and thus share headend and fibre costs between more customers. Achieving the minimum total network cost is therefore a trade-off between launch power, splitting ratio and receiver sensitivity.

Table 10.1 gives measured sensitivities for three currently available types of receiver. They relate to an SCM system carrying 32 channels with a recovered SNR (weighted) of 48 dB.

Table 10.1 Measured sensitivity for available receivers.

Receiver type	*Sensitivity*
Ge PIN	−24 dBm
Ge APD	−28 dBm
III-V SAM APD	−32 dBm

Recent work at BTL has achieved an optical budget of 34 dB using a launch power of +2 dB and a selective absorption and multiplication (SAM) III-V APD. These APDs are currently about an order of magnitude more expensive than germanium (Ge) APDs although this can be expected to reduce rapidly. At the other end of the scale, III/V PIN diodes are relatively cheap but the 8 dB lower sensitivity would reduce the passive splitting ratio by a factor of four.

A practical implementation of the network configuration of Fig. 10.3 has been demonstrated as part of the local loop optical field trials (LLOFT) in Bishop's Stortford. In these trials, which ran until the end of 1991, a 16-channel BPON system (operating at 1515 nm) is used to upgrade a TPON system which occupies the 1300 nm window. Using a laser with an optical launch power of −3 dBm, and customer receivers utilizing Ge avalanche photodiodes (APD), an optical budget of 22 dB could be achieved. Allowing for a WDM device at the customer end of the network to separate TPON and BPON, and with other losses (connectors, splices, fibre, operating margin, etc) taken account of, a splitting ratio of 16 was achieved.

The BPON system demonstrated in Bishop's Stortford represented what could be achieved at that time with available (and reasonably priced) components. An optimum configuration in terms of total network costs is the subject of ongoing study, but as technology matures is likely to be continuously changing and accompanied by falling costs.

As a technique for implementing a BPON, SCM has a number of advantages.

- Optical technology is available now and new components will improve performance as they become available.
- Existing (and low-cost) satellite reception equipment can be used.

- The use of 'radio' techniques allows for a simple interface between headend satellite receivers and the network. It also promises a seamless interface with microwave distribution systems such as M^3VDS [2].
- Analogue techniques predominate in TV broadcasting and an analogue signal is required for customers' TV sets.
- New services (e.g. MAC and HDTV) can be added simply by allocating an appropriate portion of the available frequency spectrum.

10.5.4 Digital BPON

Before describing a digital implementation in detail, the relative merits of digital transmission for TV distribution will be considered.

- Video performance in a digital implementation is determined by the encoding process rather than transmission impairments. With analogue SCM received picture quality varies across the network dependent on optical loss and may also be subject to variations with temperature and time. Provided that BER in a digital network is kept low (although video is very tolerant of some degradation) then every customer could expect the same unvarying quality.
- Digital transmission allows for easy and transparent encryption. In the technique described later, scrambling is an inherent part of the transmission and channel-selection process, so conditional access can be built in. Encryption using analogue techniques is difficult, rarely transparent and currently would require different decoding equipment for each channel.
- Studio equipment is moving towards a digital standard thus allowing source material of very high quality to be prepared. With a digital network this could be passed on to the customer without further degradation.
- Consumer video equipment is also becoming increasingly digital and making available low-cost components such as codecs. Modern TVs, cameras, VCRs and video disc players make extensive use of digital processing internally. Better performance would be achieved if digital I/O ports could be supported by digital transmission to customers.
- Advances in codec technology (lower bit rate) will allow transmission capacity or quality to be increased. Once an end-to-end digital service is established, new terminal equipment can be added without changing the network.

- Optical transmitter and receiver requirements are less stringent for digital transmission.
- Integrated circuit technology may be used extensively to reduce costs.

None of the above advantages is tenable unless a digital network can be realized at a cost which customers can afford and one which compares favourably with an analogue network. Certainly, digital video transmission would be an expensive option for the customer, if currently available trunk system demultiplexers and video decoding equipment were used. However, this is hardly surprising since they have never been designed with the severe cost restraints that application in the local loop imposes. The digital technique which will now be described is aimed at meeting the particular functional requirements for cable TV and tackling the problem of cost. In trunk systems, it is usual at the receiver to demultiplex all incoming channels. In the local loop, however, each customer would only need access to perhaps two or three channels at any time. What is required is a receiver system analogous to a TV tuner (or satellite set-top unit in SCM) which can select and decode the required channels simply and cheaply.

The time domain equivalent of a UHF TV tuner is a sampler which regularly samples the input TDM to extract a baseband channel. In this case, a bit-interleaved multiplex is required so that the sampler operates periodically at the baseband rate. A 'D'-type bistable forms an almost ideal binary sampler as it is triggered by a clock transition rather than by a narrow pulse and thus performs accurate sampling with a relatively slow clock. Furthermore it is able to perform both sample and hold functions in a single device and so fills in the time between samples. 'Tuning' or channel selection is achieved by adjusting the phase of the baseband clock relative to the TDM signal. Conventional demultiplexers rely upon a frame structure and associated synchronizing signal to permit sequential channel identification within the frame. This approach requires processing at the multiplex rate which is both technically difficult and inflexible. A neater approach, suggested here, is to transmit the channel identifier along with each channel. This allows channel selection at the baseband rate and allows variation in the choice of multiplex rate.

Figures 10.4 and 10.5 illustrate the terminal ends of a system demonstrated in the laboratory [3]. The headend (Fig. 10.4) comprises video sources, encoders, scramblers, and multiplexers. The choice of encoding scheme is determined by the baseband transmission rate, the cost of the decoder and the required picture quality. The target performance is 54 dB weighted video SNR. In the laboratory system experiment each video source was 6-bit PCM encoded to 69 Mbit/s. To improve upon cost, quality and reduce the transmission rate, a 57 Mbit/s DPCM codec has been developed using field-programmable gate array technology for the codec IC [4]. After encoding, each channel is scrambled with a unique pseudo-random sequence. As well as providing an in-built means of preventing unauthorized access to channels, this scrambling also unambiguously identifies each channel and this is used for channel selection at the customer end of the network. It also allows very efficient use of the digital system capacity, maximizing the number of channels which may be transmitted and removing the need for separate synchronization time slots. It has been found that the absence of scrambling can cause pattern dependency in the clock recovery mechanism and make the channel selection process unreliable. A synchronous multiplex of 32 scrambled signals is formed using a combination of silicon and GaAs integrated circuits and the resultant 2.2 Gbit/s stream drives a laser transmitter via a bias network. A single fibre then feeds a number of customers via optical splitters and each customer receives the synchronous TDM broadcast from the transmitter.

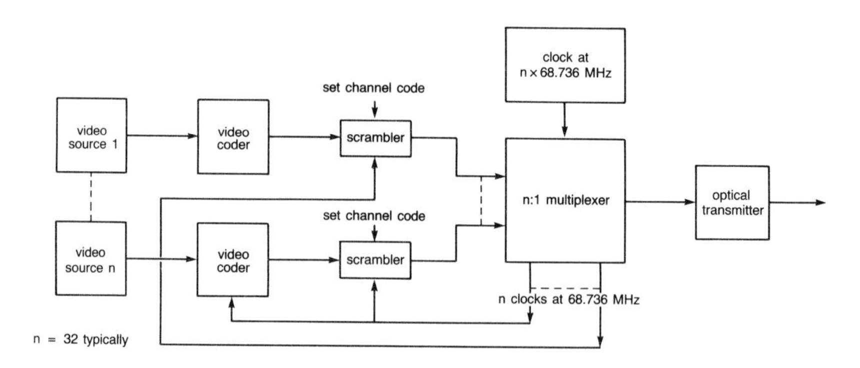

Fig. 10.4 Headend of experimental digital BPON.

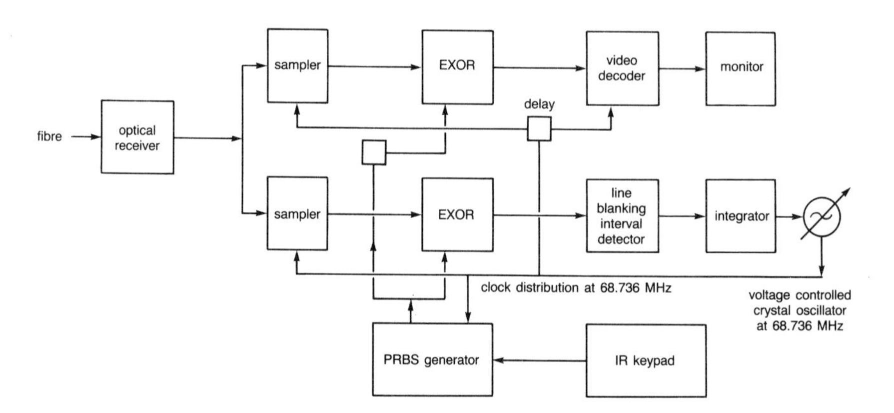

Fig. 10.5 Customer equipment in experimental BPON.

At the customer end of the network (Fig. 10.5) the only equipment working at the 2.2. Gbit/s line rate is the optical receiver. A number of receiver design options are under investigation. The lowest cost option using a PIN diode driving a separate low impedance amplifier has a sensitivity of −24 dBm at 2.2 Gbit/s. The best performance to date is a sensitivity of −34 dBm at 2.2 Gbit/s using a III/V APD. As with the SCM system described earlier, there is a cost balance to be achieved between increased receiver sensitivity (and higher cost) and higher splitting ratio which generates lower cost owing to fibre sharing.

After the optical receiver, the incoming TDM stream is split into two paths. In one path, a delay-lock loop extracts a baseband clock at 69 MHz which is phase-locked to a selected time slot in the incoming data. In the other path, the incoming signal is sampled by the extracted clock to achieve error-free data at the baseband rate. This data is descrambled, undergoes digital-to-analogue conversion in the decoder and is then passed to the TV monitor for viewing. The delay-lock loop achieves lock when the state of both the local pseudo-random bit sequence (PRBS) and the sample position in the multiplex correspond with the selected channel in the input multiplex. Thus when a new channel is selected, the local VCO slips by many cycles as the local PRBS comes into alignment with one of the headend PRBS generators. The slippage rate has to be slow enough for the line-blanking detector to operate. Thus the sampler should dwell within each channel time slot for at least 64 μs for PAL. The rate of slippage determines the mean channel selection time which is less than 0.1 s.

The system described thus far is based on a laboratory demonstrator which pointed the way to an economically viable solution. Further work has resulted in a demonstration system which aims to show that the techniques can be transferred into practical and deployable equipment with the flexibility to transport various TV formats (PAL, MAC, HDTV) plus high-quality stereo audio. In principle, this operates as already described, although during channel selection the phase of the sample clock steps through each time slot rather than slipping smoothly. This offers the potential for counting slots from the currently selected channel, thus reducing selection times. Also, a 'bit stealing' technique is used to insert data into each digitized video channel so that hi-fi audio can be carried together with a fixed bit pattern which is used in channel selection rather than detecting line blanking signals.

A further consequence of using a bit-interleaved TDM is that the potential exists for incorporating channels which require different bit rates (providing that they are still submultiples of the same overall rate). The different rate channels could then be extracted simply by changing the sampling rate in the receiver. To illustrate this, a 2.4 Gbit/s TDM system has been demonstrated which is assembled from eight 300 Mbit/s (nominal) channels.

Each of these channels can then be subdivided into individual channels in various ways. The current demonstrator uses three 100 Mbit/s channels to carry linearly encoded PAL video, but 2×150 Mbit/s might be used to carry MAC or compressed HDTV. Equally, lower bit-rate codecs could be used to increase channel capacity (e.g. 4×75 Mbit/s giving 32 channels total).

In addition, further laboratory work is aimed at exploring the potential of this technique in terms of increased bit rate and to investigate the use of optical sampling rather than the electronic sampling described earlier. Transmission at 4.4 Gbit/s with a receiver sensitivity of −21 dBm has been demonstrated [5], as has the principle of optical sampling at 2.2 Gbit/s [6] which gave the advantage of using an optical receiver with a bandwidth of only 56 MHz compared with over 1 GHz when using electronic sampling at this rate.

10.6 CONCLUSIONS

This chapter has examined how a passive optical network could be upgraded to incorporate broadband services. Two particular implementations have been studied, one analogue the other digital, each of which has its own advantages and drawbacks. Looking at optical performance and the degree of splitting possible neither technique shows a clear advantage, although in considering video performance a digital network would have the edge, not just in terms of its ability to meet a particular specification but also in the confidence level of maintaining that performance.

The dominant cost factor in either system is the optical receiver required by each customer, but since the technical requirements are very similar this again is not a deciding factor.

However, analogue subcarrier multiplexing is a 'here and now' technology and would certainly allow earlier deployment. Longer term, the goal of network studies worldwide is digital transmission in the local loop and the systems described here have indicated a cost-effective way forward. Certainly it has been shown that a digital system can be designed to provide the functional requirements and flexibility so readily provided by analogue systems. The question still to be answered, and one which is the subject of ongoing work, is when will a digital network be cost-effective, bearing in mind that the technology available for SCM will be continually improving and reducing in cost?

Particular technological developments which will have enormous bearing on the local network and the provision of services are high-density WDM and erbium-doped fibre amplifiers. Erbium amplifiers offer the prospect for increasing optical splitting ratios by orders of magnitude (even perhaps for

AM VSB systems), whilst WDM has the potential to make available enormous bandwidths, with sufficient capacity for each customer to have his own dedicated wavelength. Despite these technological changes, however, the clear advantage that the passive optical network has over other approaches is one of durability. New developments will serve to enhance the network in an evolutionary way rather than limit its useful life as often happens with advances in technology.

REFERENCES

1. Ritchie W K: 'The BT switched-star network for cable TV', BT Technol J, 5, No 4, pp 5-17 (September 1984).

2. Pilgrim M, Scott R P I, Carver R D and Ellis B J: 'The M^3VDS Saxmundham demonstrator — multichannel distribution by mm-waves', BT Technol J, 7, No 1, pp 5-19 (January 1989).

3. Faulkner D W, Russ D M, Douglas D and Smith P J: 'Novel sampling technique for digital video demultiplexing, descrambling and channel selection', IEE Electronics Letters, 24, No 11, pp 654-656 (May 1988).

4. Faulkner D W, Cook A R J, Bunting P: 'Digital TV distribution in the local loop', Proc EFOC-LAN 90 (June 1990).

5. Smith P J, Lobbett R A and Faulkner D W: '64 channel digital TV distribution system operating at 4.4 Gbit/s', IEE Electronic Letters, 24, No 21, pp 1336-1338 (October 1988).

6. Faulkner D W et al: 'Optical samplers for gigabit cable TV receivers', IEE Electronic Letters, 24, No 23, pp 1430-1431 (November 1988).

11

COMPONENT DESIGN FOR PASSIVE OPTICAL NETWORKS

S Hornung

11.1 INTRODUCTION

A single-star optical network is used in some high-traffic situations for provision of telephony in the access network [1]. Each customer has an individual optical link to the central access node and there is no need for active electronics in the external part of the network. It thus forms the most basic 'passive' network (see Chapters 4 and 9).

Such systems typically provide four fibres to each customer in a direct point-to-point manner, with a fibre for each direction of transmission and a back-up system running in hot standby. This network topology results in high fibre-count cables and a cost of provision that can only be justified for large business users.

This chapter describes optical networks designed to meet the 'passive' aim of no street-mounted electronics (or at least to evolve smoothly to this goal), whilst being geared more in cost to the small business or residential user. Cost reduction is obtained by using the fibre bandwidth to share one fibre amongst a number of customers, reducing the amount of plant in the ground and also sharing the exchange equipment.

One particular approach using optical splitters will be described in detail, including optical design, critical components and how such a network might be tested and maintained. An alternative approach using wavelength multiplexing will be briefly described. Finally it will be shown that careful design of the plant can allow either network to evolve easily from networks that use street electronics and, perhaps more importantly, can evolve easily to point-to-point networks by simple replacement of components as service demand grows.

11.2 ADVANTAGES OF FIBRE PROVISION

A well-designed fibre scheme will offer certain advantages for telephony provision.

- Rapid provision of extra lines — the bandwidth of the fibre is such that multiplexing schemes can be used that allow flexible provision and reconfiguration of telephony service.
- The ability to 'fully manage' the network — this means the ability to monitor for faults continuously using supervisory channels and in some cases the ability to route around them. Again the extra bandwidth of the fibre is the enabling feature.

In addition the schemes outlined in this paper have some extra advantages as detailed below.

- The ability to provide additional services on alternative wavelengths — an obvious example here is the provision of broadcast TV to residential customers, while business customers could use the extra wavelength for LAN interconnect, etc.
- Reduction of the amount of plant in the ground and the sharing of exchange equipment — more specifically, these schemes could remove the need for high fibre-count cables and reduce space requirements in the exchange.

11.3 PASSIVE NETWORK OUTLINE

In a major access network, such as the one operated by BT, the existing copper network is based on a topology having two flexibility points — at streetside cabinets serving up to 600 lines and at distribution points (DPs) serving around 10-15 lines. The duct network tends to reflect this topology.

The difficulty and cost of relocating duct work in practice mean that future optical schemes must accommodate to the existing plant layout.

The first approach makes use of passive splitters and is shown in Fig. 11.1.

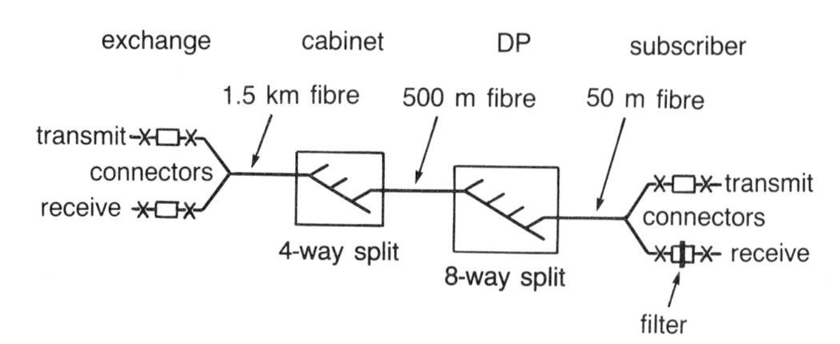

Fig. 11.1 Basic passive splitter network with average lengths.

In this network a single fibre from the exchange is fanned out via passive splitters at both cabinet and DP points to feed a number of individual customers. A time division multiplexed (TDM) signal is broadcast to all terminals from the exchange on a single wavelength, with customers time-accessing the particular bits meant for them. In the return direction data from the customer is inserted at a predetermined time to arrive at the exchange in synchronism with other customers' data. Inclusion of an optical filter in the customer's terminal that passes only the telephony wavelengths without disturbing the telephony wavelength allows the later provision of new services on other wavelengths without disturbing the telephony transmission. A target of a 32-way split operating at 20 Mbit/s will allow the provision of basic-rate ISDN to all customers. Table 11.1 shows a comparison of the amount of plant per customer for various split levels of such a network compared with the amount of plant needed to feed a single fibre to each customer. It can be seen that the amount of fibre per customer is dramatically reduced for the split network. Some saving results for splices and connectors. Surprisingly, the split network has the same number of fibre splitters per customer as the single-fibre feed owing to the sharing of splitters towards the exchange. It should be noted that the single-fibre feed requires a splitter on each end to route both directions of transmission down the same fibre — this is known as bidirectional or duplex working.

Table 11.1 Plant per customer for the network of Fig. 11.1.

Plant per customer	*Fibre (m)*	*Splices*	*Connectors*	*Splitters*
Point to point duplex	2050	6.00	4.00	2
TPON 32-way split	159	4.34	2.06	2
TPON 16-way split	206	4.44	2.13	2
TPON 8-way split	300	4.63	2.25	2

The second approach is to give each customer a separate wavelength channel on a shared fibre and by means of wavelength division multiplexers (WDMs) create a virtual single-star network. Wavelength division multiplexers are used at each end of the fibre to combine several wavelengths on to the fibre and separate them at the far end. By mounting one of the multiplexers at the exchange and the other at the DP, the plant between the exchange and DP can be shared by the number of wavelength channels. Normally this approach requires a separate wavelength laser source for each channel. An alternative approach avoids this by using edge-emitting light-emitting diode (ELED) sources for each channel having the same wavelength output. Each channel has an ELED source but all the sources are the same and have a wavelength spread wider than the total spread of the WDM channels. The first WDM component then selects a different wavelength 'slice' from each ELED, which is then directed to the appropriate output by the second WDM. This spectral slicing approach is shown schematically in Fig. 11.2. A 10-channel system has been demonstrated in the laboratory.

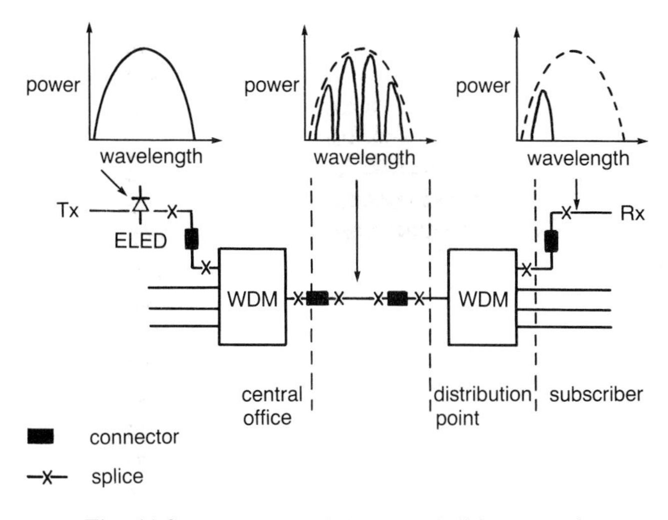

Fig. 11.2 Schematic of the spectral slicing network.

Although differing greatly in detail, the splitter and WDM approaches have much in common from the viewpoint of network design, installation and testing. Key points can be summarized as follows.

- Plant in the ground is reduced by the shared use of fibre.
- The fibre is required to carry a number of wavelengths, carrying different channels or services. Some 'spectrum management' is required and unwanted wavelengths must be filtered out at the receive points.

- Optical components such as splitters and wavelength multiplexers that have hitherto not been used in the field must be made low-cost and rugged enough to be easily and reliably installed and used in the external part of the network.
- The shared nature of the networks means that the fibre cannot be disconnected to test for a fault in one customer's service without disconnecting other customers. Ideally the networks should be tested 'live' without disrupting service.
- If bidirectional working is used (as in Fig. 11.1) or reflection-sensitive systems (for example microwave subcarrier or coherent systems) are contemplated for future use, then attention will need to be paid to reflection levels from splices, connectors and unused fibre outlets on splitter arrays. In this chapter this is explored for the case of bidirectional working.
- Not specific to the particular architecture is a need for a cable and housing infrastructure that can effectively route fibre through cabinet and DP positions and into and around business and residential premises whilst providing adequate fibre management and testing facilities. Included here is a need to house and power the customer's equipment.
- Overlaid on the above factors is a general need to keep component costs low and installation practices rapid and simple if cost targets are to be met and existing local network staff used effectively — this is of utmost importance.

In the next section of this chapter the work undertaken to solve these problems for the passive splitter approach is described in detail, including the description of a working demonstrator and of a statistical model designed to provide worst-case and dynamic range information on network loss.

11.4 THE PASSIVE SPLITTER NETWORK

11.4.1 Network outline

The basic passive splitter network has been shown in Fig. 11.1. A 32-way split network is used in the field trials. To fit in with the cabinet/DP geography of the network, this is planned as a four-way split at the cabinet point and an eight-way split at the DP point. A full schematic of this network is shown in Fig. 11.3. In some instances a topology employing a combination of a star splitter with a distributed splitter provides the most suitable fit to the existing network as shown in Fig. 11.3.

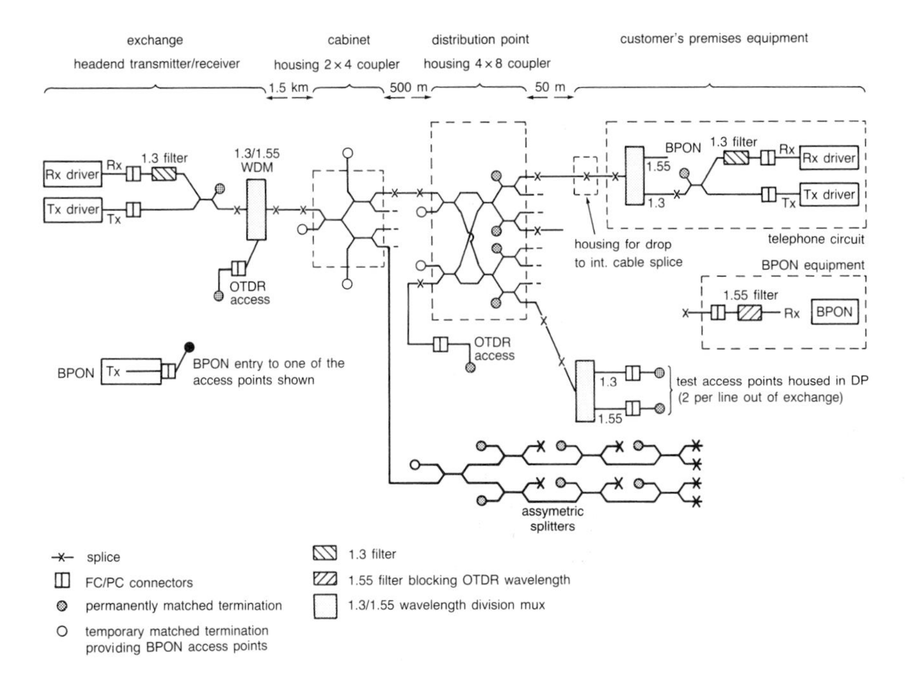

Fig. 11.3 Full schematic of the passive splitter network.

The operation for telephony has already been described in outline and is covered in detail by Dufour [1]. Of particular importance is the need for the power levels received at the exchange receiver to exhibit as little bit-to-bit variation as possible. This is covered further in a later section.

11.4.2 Wavelength allocation

The passive splitter-based network is designed to carry a range of services, which may be on separate wavelengths. These are introduced on to and taken off the network by wavelength division multiplexing (WDM) components. Reception of unwanted wavelength channels is prevented by using bandpass filters in front of the detectors.

Each service has to be allocated a part of the optical spectrum bearing in mind that there will be a finite width to the laser spectrum as well as an uncertainty in its position (due to production tolerances and temperature shifts). Tolerances in width and wavelength position of the matching filters must also be taken into account. Other characteristics which must be considered are the spectral shapes of WDMs used and the spectral shape of the system transmission loss including the fibre 'water' (OH) absorption peak.

The customer end laser is the dominant component cost. To allow the use of low-cost lasers, the entire 1.3 μm window between the ranges 1.26 μm and 1.34 μm is assigned to TPON (telephony). The 1.575 μm region is assigned to optical maintenance whilst the 1.500-1.550 μm region is available for broadband services. The broadband region could later be divided further by high-density WDM.

The 1.3 μm region and the 1.55 μm region can be separated by a fused-taper WDM.

In addition all of the installed plant must be wavelength 'insensitive' or low loss across a range of wavelengths. In the case of the splitter arrays, this includes the need for constant coupling ratios across the required wavelength band to avoid wavelength 'steering' by the network.

11.4.3 Critical component needs

From the preceding discussions the components or factors important to the passive splitter-based network can be summarized as follows:

- wavelength-independent splitter arrays;
- blocking filters to exclude later wavelengths from the customer's receiver;
- lasers and associated electronics in the customer's premises that can be remotely controlled in output level to equalize levels received at the exchange;
- low fibre-count cables for cabinet and DP interconnect and customer's drop (overhead and underground) and internal applications;
- housings for splitter arrays and customer's equipment with adequate fibre management and testing facilities;
- testing and monitoring facilities that can be employed with the network 'live';
- consideration of reflection levels from splices, connectors and unused splitter outlets;
- low-cost components easily used by local network staff.

11.4.3.1 Splitter arrays

The heart of the optical network is the splitter, an efficient device capable of dividing the arriving optical power between many ports with minimal excess loss.

There are several technologies capable of producing the necessary devices. Fused biconic couplers are made by heating and drawing two optical fibres in contact [2]. Light launched into one will couple into the other, giving 50% coupling. Several can be cascaded together to give an array with a large number of output ports. If required, the coupling ratio can be changed to meet the fixed needs. Larger groups of fibres may also be fused and drawn, giving 1×7 and even 1×19 splitters.

Splitter arrays are now also produced by a planar glass waveguide technology [3]. Devices are available for 1×2, 1×4, 1×8 and 1×16. It is envisaged that these devices, once in mass production, will be low-cost.

The performance of the optical splitter is determined by its excess loss, the accuracy of the 50% power split and, since the network is required to be broadband, the deviation from the 50% power split over a wide wavelength range. Figure 11.4 shows a typical measurement.

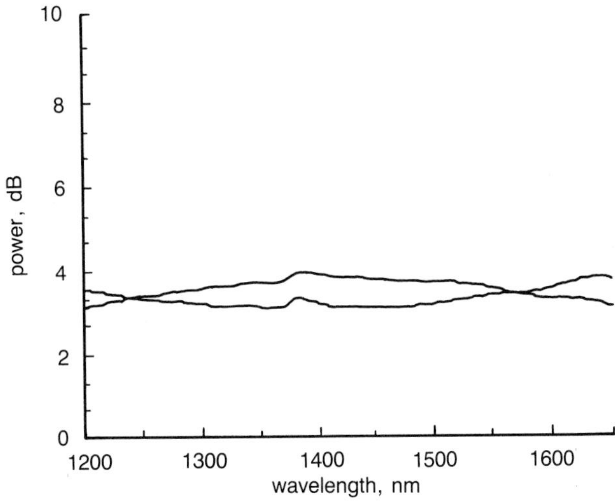

Fig. 11.4 Wavelength response for a single splitter.

Fused 1×2 biconic couplers have typical excess losses of 0.05 dB, are within 2% of 50% and are wavelength-independent within a 0.5 dB band over the range from 1.260 μm to 1.600 μm. In addition, the coupling ratio is independent of input polarization to 2% and the performance is maintained between − 40 °C and + 70 °C.

The planar waveguide equivalent has an overall throughput loss less than 4 dB for the wavelength region of interest. The polarisation and temperature dependence of this device is negligible.

The fused taper splitter has now been available for some time and therefore has good optical performance. The planar splitter has emerged relatively recently and offers the prospect of mass production and consequently lower cost.

11.4.3.2 Optical filters

A bandpass optical filter is used at the telephony customer's receiver (and at the headend), to block all signals other than telephony. An example of such a device employs a multilayer dielectric coating on a thin (100 μm) silicon wafer.

The important performance parameters are the excess loss and the out-of-band rejection. Filters with excess losses below 0.4 dB and out-of-band rejection of 25 dB have been used.

The 1 mm square filters are mounted in a precision slot cut through a fibre connector ferrule, with fibres inserted from both ends. The filter is 'sandwiched' between the fibre ends, with a total fibre-to-fibre loss of below 1.5 dB. The construction is shown in Fig. 11.5. The filter could also be incorporated into the receiver itself or it could be provided by a length of doped fibre [4].

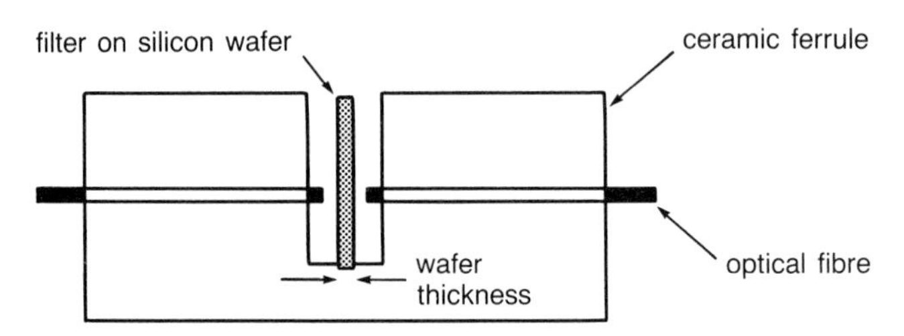

Fig. 11.5 Prototype optical blocking filter.

11.4.3.3 Optical transmitters and receivers

In the exchange-to-customer direction the telephony provision on the passive splitter network is a fairly conventional digital transmission system at, typically, 20 Mbit/s. However, in the customer-to-exchange direction a large number of laser transmitters are interfaced to one receiver. Since each customer path could experience a different loss owing to component or path length variations (see section 11.4.5) it is likely that the amplitude of bits received at the exchange could vary widely from bit to bit unless steps are taken to control the output of customer's lasers.

These variations could be up to 15 dB for the current network design, necessitating laser transmitter designs that can be turned down by this amount from a nominal level (−3 dBm in this design) and still operate correctly at 20 Mbit/s. Studies of laser noise, spectrum and turn-on delay under these conditions indicate that no fundamental problems are likely to be encountered for typical 1300 nm lasers. A penalty of around 1-2 dB is likely at the exchange receiver from residual (1 dB) bit-to-bit variations.

In order to reduce costs, it is likely that initial implementations to the passive splitter networks will use conventional Fabry-Perot (FP) type lasers for telephony use, in the 1300 nm window. The production tolerance and temperature variation of these lasers is such that the whole of the 1300 nm window must be reserved for this one service.

For broadband use in the 1550 nm window, both dispersion effects and the possible need to have several wavelengths (for different services) dictate the use of lasers with a tighter spectral control, such as distributed feedback lasers (DFBs). Although much higher in cost at the moment they are predicted to reduce to the cost levels of the FP lasers with volume and time.

11.4.3.4 Cables

By the nature of the splitter network, only a few single-mode optical fibres emanating from the telephone exchange are needed to serve a large number of customers. Typically, ten fibres into a street cabinet can provide service to about 300 customers on a duplex passive splitter network.

Future cable designs for local network access will reflect this efficiency and physical size will no longer be the dominant factor in duct utilization. In fact, the current limitations on minimum size can be reviewed once larger, heavier, copper cables have been removed from duct routes.

11.4.3.5 Underground cables

Low fibre-counts and small diameter cables will be the aim for shared-fibre passive networks. These cables will be lighter and more flexible; therefore longer sections between joints will be possible, leading to higher reliability and lower costs.

The blown fibre concept will also play a major role [5]. Continuous links of small fibre bundles will be blown through a network of small tubes giving splice-free circuits between the flexibility points at exchange, cabinet and distribution point (DP). The technique also allows flexible provision since extra fibres can be quickly and cheaply installed once the microduct assembly is in place.

Current optical cable specifications apply. All these specifications typically require <0.5 dB/km optical loss in both the 1.300 μm and 1.550 μm transmission windows, over a temperature range from 60 °C down to −10 °C.

11.4.3.6 Overhead cables

Much of BT's copper local network uses overhead (OH) cabling, particularly to provide the final customer feed. This provides a very cost effective flexible access facility and will require a fibre equivalent for future optical network access.

OH optical cables will require broader and more stringent specifications than their underground (UG) counterparts. Typically a temperature range from 80 °C down to −20 °C is required in the UK. Strain-resistant fibres are attractive for OH use and the blown fibre system can be configured to limit fibre strain to acceptable levels even for currently available fibre and under the most arduous conditions of ice and wind loadings.

Both conventional and blown fibre drop cables have been developed, some containing one or two copper pairs (Figs. 11.6 and 11.7). The blown fibre version has the advantage that the installation strain imposed on the structural components is not present in the fibre, which is blown through after stringing up the tubing.

The conventional OH 'figure of 8' design (Fig. 11.6) uses two fibres proof-tested to 1.6% in a loose tube with jelly fill. This enables an allowable maximum strain from environmental loading by wind and ice, of up to 0.4%.

The blown fibre OH drop cable (Fig. 11.7) is similar in design using the same strength member with five PVC-covered steel wires and one copper pair, but a 3.5 mm bore tube replaces the loose tubed fibre in the bottom lobe. Owing to the reduced strain in blown fibre resulting from blowing the fibre unit after stringing, a standard four-fibre bundle can be used for the customer feed, with 0.7% strain-proof fibre and two Kevlar rip cords for easy access.

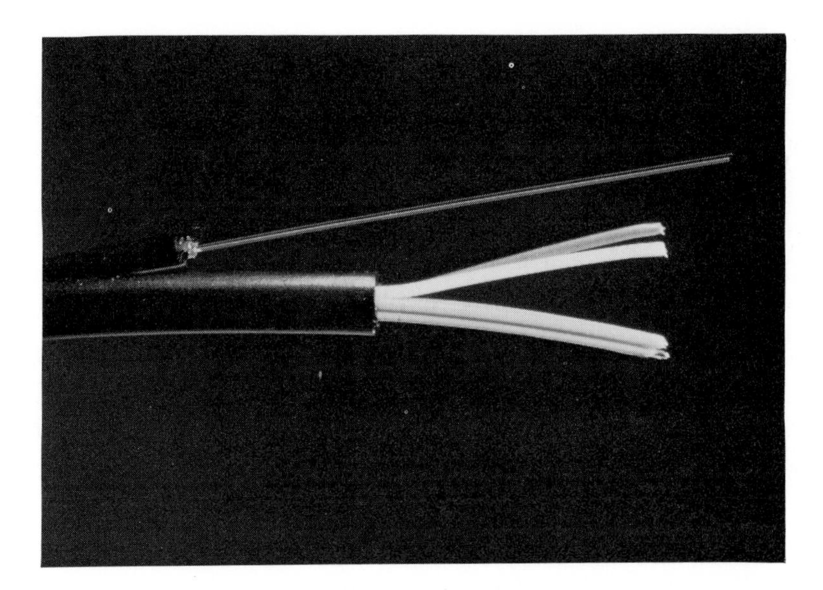

Fig. 11.6 Conventional overhead drop cable.

Fig. 11.7 'Blown fibre' overhead drop cable.

11.4.3.7 Fibre upgrade to copper cables

For the fibre-to-the-kerb system (street TPON) (see Chapter 4) the optical drop cable is not required for the first implementation phase. The copper pair is used to provide simple telephony to a normal house installation. At a later time, broadband access can be provided over the fibre when cable TV is requested.

This concept of copper now and fibre later can be extended throughout the existing network and composite cables have been considered from time to time.

There are two major drawbacks when considering fibre-copper composites. The first is the cost penalty resulting from the need to change all existing copper plant technology to new systems and techniques capable of storing and preserving the fibre tails for future use. The second is that of risk versus craft skills. In order to reduce risk, fibre skills must be taught to all cable technicians, a major investment for a national implementation policy. Even with such an investment, there is no guarantee that the fibres will be in good condition when required, as an error on initial installation will not be apparent until the fibres are needed.

A solution to these problems can be identified with the blown fibre approach. By simply providing low-cost tubes in copper cables, minimal skills are required for the cable technician to extend a sealed tube-tail to outside the copper joint closure. Subsequent access to the tubes can then be provided, without disturbing the copper joint, by cutting off the seals and fitting a secondary joint housing to the tails. The dome-ended DP closure would be a typical solution. This housing can then be used for fibre splices, passive splitting or other configurations once fibre is blown through the cables.

11.4.3.8 Housing and flexibility points

A telecommunications provider such as BT, has well-established procedures and structures in its local distribution network. A valuable asset is the trained and experienced workforce and a phased changeover to optical fibre must utilise all this existing investment. Wherever possible, it is intended that optical plant developments will integrate easily with existing structures and a modular conversion technique is envisaged.

Optical fibre requires a new philosophy in the concept of external plant. For the passive splitter network there will be far fewer fibres than there are presently copper wires. However, fibres require greater care in handling, are more difficult to join and must not be kinked or tightly bent. At present, for most telecommunications providers such as BT, fusion splicing has proved to be the most cost-effective fault-free method of joining fibres, although field termination of connectors and mechanical splices look promising for the future.

All of these techniques require fibre tails of length sufficient to reach from the cable end termination or optical component module out to a fusion splicing or connector polishing machine. Typically, these tails will be 1 m long and require controlled storage to ensure accessibility and long life in conditions of high humidity.

With passive networks employing fibre splitters, many customers will be served from one fibre and access to each fibre for testing, using non-intrusive 'clip-on' techniques, will be necessary to limit interruptions to service (see section 11.4.3.13 below).

All these factors have influenced the design concepts of developments incorporated in the modular approach to optical plant.

11.4.3.9 Fibre and splice storage

Signal transmission on any medium can be affected by variations in the attenuation of the medium. With optical fibre, this is probably the most predictable problem. Any undue bending or kinking of signal-carrying fibres can result in increased optical loss of several dB. This can easily exceed the automatic gain control or digital recovery provision of the transmission equipment.

To cater for and protect against this problem, each pair of fibre tails and the associated splice or connector have been assigned a single tray in the modular concept. A tray has been designed to cope with the majority of perceived needs in any module, and each tray has a separate, variable fibre storage capacity for each tail, to allow for re-splicing and 'clip-on' test access (Fig. 11.8). This approach gives maximum security to each fibre circuit and eliminates accidental intrusion on adjacent live fibres.

11.4.3.10 Street cabinet

The passive optical network must interface with existing external plant, so each cabinet module will consist of one splitter array and sufficient single splice trays to accommodate all the array tails and some spare circuits for later broadband upgrade. In a typical duplex 32-way split a 4×4 array will be located at the cabinet. A minimum of eight trays will be needed, therefore ten will be provided in the completed module.

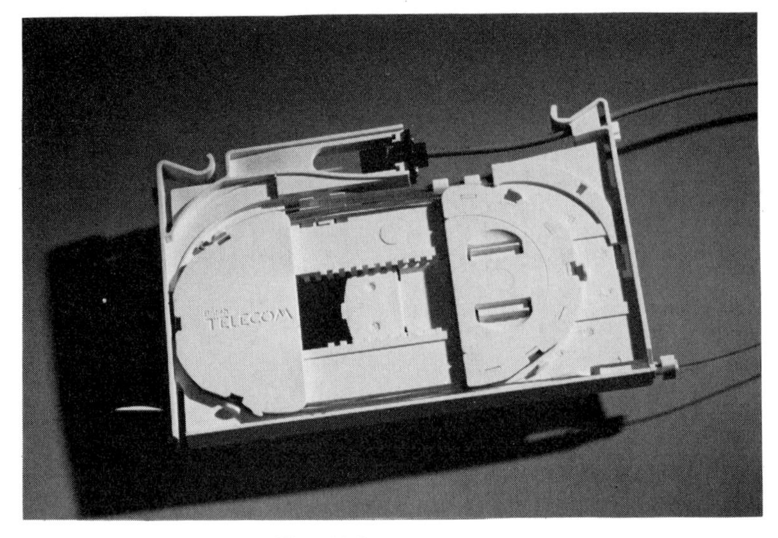

Fig. 11.8 Splice tray.

A standard cabinet, used at a flexibility point, can serve up to 300 customers in the copper network; therefore ten 4×4 array modules will be fitted on a fibre upgrade exercise, giving capacity for up to 40 DPs from the one cabinet.

11.4.3.11 Distribution point (DP)

DPs must feed customers with OH or UG drops to suit the local topology, but it is not currently possible to splice fibres at the top of a pole. A universal DP module (Fig. 11.9) housed in a 180 mm diameter cap-ended closure will be accommodated in a footway box with OH drops fed down the pole in plastic conduit. UG feeds will radiate from the box using ducted or direct buried cable as required. All customer feeds will be in the form of a continuous spliceless link from the DP to the customer's premises. This is easy to accommodate with blown fibre as tube joints and type changes can be used to simplify the installation and the fibre blown through later, in one length.

For the typical duplex 32-way optical split the DP has a single fibre feed from the cabinet with an eight-way splitter array. One leg from the array is fitted with a high-return-loss single-mode connector for OTDR test access.

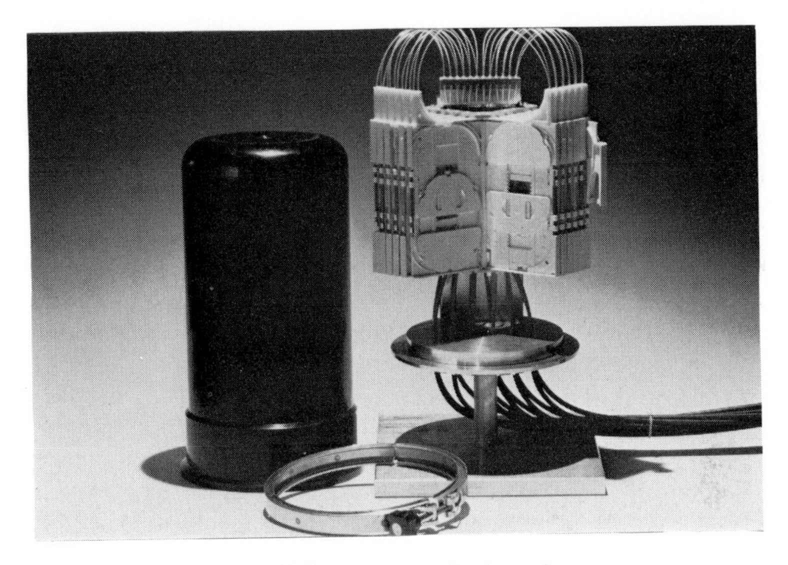

Fig. 11.9 Fibre distribution point.

Thus seven customers can be fed from the remaining legs and a total of 12 single splice trays make up the complete DP module.

11.4.3.12 Customer flexibility requirements

Arrangements for terminating fibre at the customer's premises will depend on many factors such as type of property, method of feed and required service. With any shared transmission system such as TPON, special measures must be taken to ensure privacy of the customer's information. This will be primarily achieved by electronic coding of the signals. In addition, however, direct access to the fibre will not be allowed and therefore the network-terminating equipment and the optical components and splices will be contained in tamper-proof housings.

With all feed arrangements there will generally be an internal-to-external interface. Ideally, no splice will be required here, but if the equipment is located any distance from the entry point, internal cable will need to be run and a splice housing will be required. This must also be tamper-proof and able to cope with internal and external locations. As with other flexibility problems blown fibre generally simplifies the situation as only tube changes need accommodating.

11.4.3.13 Testing and maintenance

The maintenance of the optical network falls broadly into two categories:

- installation and commissioning;
- fault location and cure.

The first ensures that the system is within specification as it is constructed and provides a database to which a maintenance engineer can later refer when trouble-shooting. Clip-on power measurement techniques may be used during network build for confidence testing and conventional end-on power meters used for end-to-end commissioning.

A novel feature of a splitter-based network is that it needs to be maintained live so that the testing of a fault on one customer does not interrupt the others. The initial maintenance strategy has three main features:

- allocation of a maintenance wavelength which can be used to check the continuity and optical performance of the network without referring to the transmission equipment, so that faults associated with the fibre can be distinguished from electronic equipment failure;
- use of non-invasive testers using the clip-on approach (shown schematically in Fig. 11.10), a small bend in the fibre being used during system operation to tap out a small proportion of the light which is picked up by a waveguide and guided into a detector for measurement;
- use of the in-built laser power control on the system to correlate power with network faults.

The splitter-based network is characterized after installation in terms of insertion loss from the headend to each customer termination at each working wavelength and at the maintenance wavelength. Return (reflection) loss is also measured.

Optical time domain reflectometer (OTDR) traces are recorded in two stages, from headend to the DP and from the DP to the customer's terminal. Measurement on the network must be split in this way because measurements made through a splitter are reduced in resolution. For example, the OTDR measurement of a point loss at the output of a four-way splitter would only increase by 25% if one of the four fibres broke. In addition, the OTDR alone is not capable of distinguishing individual paths beyond the splitter. The trace is, however, useful for identifying the exact location of the fault and any change in performance. These are then correlated with any known faults in

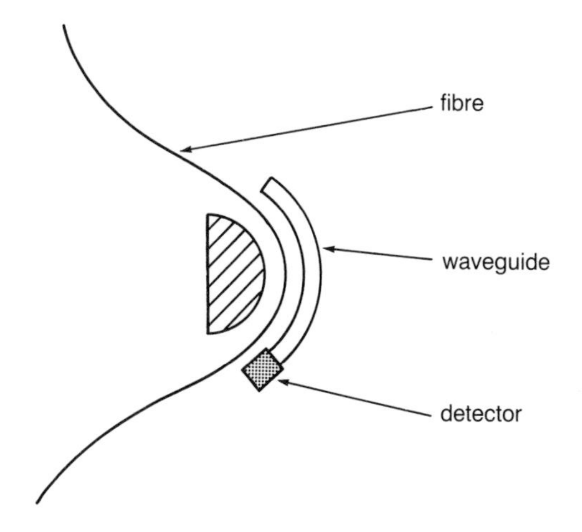

Fig. 11.10 Schematic of optical 'clip-on' testing.

one or more customer's equipment to enable a unique path and location of the fault source to be identified.

'Clip-on' power meters calibrated at 1.575 μm and incorporating a band pass optical fibre at this wavelength can be used to make measurements around the location of the source of the fault to identify the faulty component.

In the long term, continuous monitoring equipment may be built into the headend, which has a constant, up-to-date picture of the status of the optical network allowing rapid response to faults or demands placed on the network.

11.4.4 Splicing and connectorization

Single-mode fibres must be aligned to a precision of less than 1.5 μm to achieve joint losses of less than 0.5 dB. Low-cost, easy-to-use techniques are required for both permanent (splicing) and temporary (connectorized) jointing. In addition to the loss requirement, if splices or connectors are to be used in the bidirectional part of the passive splitter network, they must have low reflection levels since reflections appear at the near-end receiver as crosstalk. Levels of return loss around 50 dB are required (i.e. reflections must be 50 dB below incident power).

The type of connection method used is dependent on its location within the passive network. Currently, fusion splicing is the only technique capable of reliably meeting the loss and reflection performance required for the external part of the network. The technique is effective and widely used but needs an expensive machine and hence is a relatively high-cost termination method. This can be reduced through the use of a multifibre splicing process which has the potential to reduce splicing time by up to 75%. However, fusion splicing will remain impractical for some network requirements (e.g. for test access) and other jointing techniques have to be used.

Connectors will be used where they are essential, for example in the termination of fibre on to exchange or customer equipment and at test access points. For network terminations conventional connector styles (e.g. the Japanese FC/PC) can be used. However, to reduce costs, connectors with moulded outer shells and quicker production times (e.g. the Japanese SC style) are attractive. For test access applications low-return-loss connectors are required. One example currently under field evaluation is the ST style with a ferrule face angled at 6° which gives a return loss of >50 dB in both the unmated and mated state. This technique of angle polishing the ferrule face has been proved for other connector designs and multifibre connector types.

Alternative splicing methods typically employ mechanical alignments such as precise grooves, with UV-light cured glues used to hold the fibre in place. Such mechanical splices normally have large reflection levels due to poor matching of the glue refractive index to that of the fibre, but can be extremely cheap and easily terminated in the field with low-cost tools. New prototype glues have achieved the required matching to give return losses >50 dB. Other techniques have included polishing the fibre faces at an angle of 8° before assembly. One possible application of mechanical splices is in situations where fusion splicing is impractical or would result in a high termination cost. For example it can be used to terminate factory-polished fibre pigtails to network fibre in customer premises and as a temporary repair for damaged cables.

In addition to methods of jointing individual fibres, some potentially low cost multifibre connectors are beginning to emerge. If low reflectivity versions that can be terminated in the field become available then they could be used to connect the splitter arrays into the network, leading to the possibility of rapid replacement of arrays for maintenance purposes and to the evolution possibilities discussed later.

11.4.5 System model

The splitter-based optical network contains a range of optical components of different performance. In order to be able to predict the overall network performance from individual components a system model is required; components can then be specified against the needs of the overall system.

Representative statistical data has been acquired of all passive components which make up the splitter-based network including splitters, optical connectors, optical fibre spices, WDMs and filters. The data gives the range of optical parameters that may be expected, and was used to calculate the three sigma loss points of a 32-way network at 1.3 μm to be 35.2 and 24.0 dB. These figures therefore represent the 'worst case' loss of the longest route containing poor components and the 'best case' loss of a short route containing good components at a single wavelength.

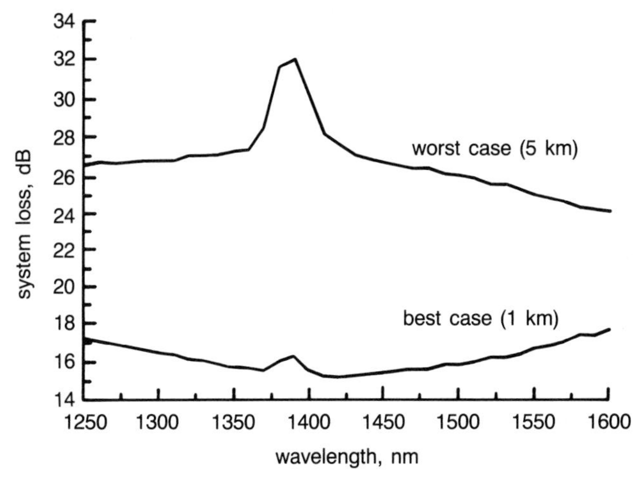

Fig. 11.11 End-to-end worst- and best-case system model loss prediction.

Taking the complete wavelength spectra for each component, the 'best case' and 'worst case' spectra (Fig. 11.11) have been calculated which define the working range over the entire usable spectrum from 1.25 μm to 1.6 μm. It is clear that the OH absorption blocks the range of 1.36-1.42 μm. The spectrum does not include contribution from wavelength-filtering components such as WDMs and filters, and represents the fibre/splitter network only. The wavelength-filtering components define the spectral allocation within which particular systems, including maintenance, are allowed to operate, as discussed in section 11.4.2.

11.5 DEMONSTRATION SYSTEM

In order to test the various trade-offs inherent in the passive splitter network a demonstrator system was built, as shown in Fig. 11.12. Two exchange points and two DPs were interconnected via a passive splitter array mounted in

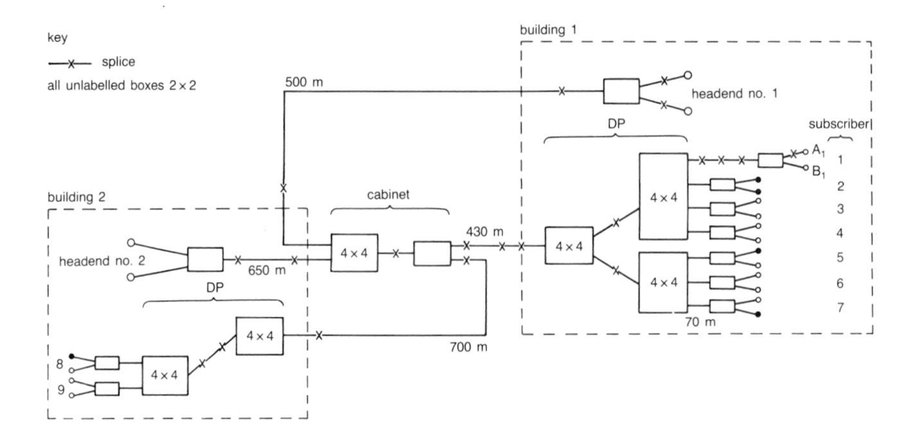

Fig. 11.12 Schematic of the passive network demonstrator.

a standard external cabinet to simulate a 32-way split. The total system length was around 1.5 km, with blown-fibre cable being used throughout.

The measured loss of the system at 1.3 μm was 35 dB, in comparison to a loss of 34 dB predicted by the system model.

As previously described the system could be tested with clip-on testers or by means of OTDR equipment launched via a low-reflection connector test access point in the external cabinet.

11.6 EVOLUTION

Thus far the passive splitter network described has taken fibre all the way to the customer's premises with splits at cabinet and DP. However, initial deployment may be more economically achieved by stopping the fibre network at a street-mounted multiplexer, retaining the existing copper pair for the drop to the customer (see Chapter 9). At the other end of a usage/traffic spectrum the large business customers require one or more fibres each.

Thus the pure passive network might be seen as sitting in the middle of an evolutionary path, with street-mounted electronics at the start and one fibre to each customer as the final end point. To further complicate matters the middle ground has several alternatives (for example the wavelength network described in section 11.3).

Can the network plant and structure be designed to accommodate and cater for the evolution and alternatives?

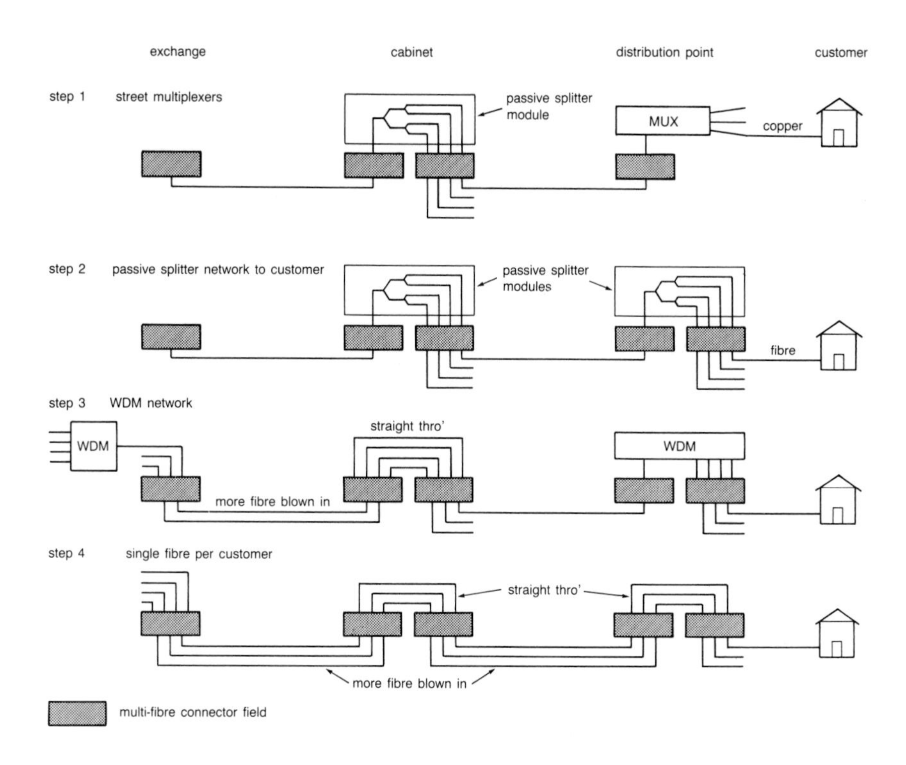

Fig. 11.13 Evolution of passive networks.

The development of blown-fibre cable and field-mounted multifibre connectors gives the key to one possible solution. Assuming that all components are connected with multifibre connectors and blown fibre, Fig. 11.13 shows a possible way in which a common family of parts can enable the network to start with street multiplexer options and progress through passive splitter or WDM networks to single fibre per customer by simple replacement of components at the multifibre connector points and/or the addition of more fibre by blowing into existing tubes.

11.7 CONCLUSION

In this chapter some possible options for the local fibre network of the future have been described, based on fibre-sharing techniques to reduce cost. One particular option using passive splitters has been described in detail, including network and component design, testing and maintenance.

Optical fibre systems open up enormously exciting prospects for service provision and integrated equipment design in the access network, but this must not blind us to the practical problems of introducing and maintaining these networks. It can be seen that the components and work practices are already here in prototype form; the challenge is to advance the technology rapidly to the realisation of a low-cost, high-quality, high-service-capability network. Fibre technology has risen to this challenge in the core network; it surely can in the access network — its next major challenge.

REFERENCES

1. Dufour I G: 'Flexible access systems', BT Eng, J, 7, No 4, pp 233-236 (January 1989).
2. Mortimore D B: 'Wavelength flattened 8 × 8 single-mode star coupler', Elec Lett, 22, No 22, pp 1205-1206 (1986).
3. McCourt M et al: 'Optical and environmental performance of packaged single mode I × N couplers made by ion exchange in glass', OFC '90, paper WE1 (1990).
4. Wilkinson I J, Finegan T and Ainslie B J: 'Application of rare-earth doped fibres as low pass filters in passive optical networks', Elect Lett, 27, No 3, pp 230-232 (1991).
5. Hornung S, Cassidy S A, Yennadhiou P and Reeve M H: 'The blown fibre cable', IEEE J on Sel Areas in Telecom, SAC-4, No 5, pp 679-685 (August 1986).

Part Four

Switching, Signalling and Operational Support

12

THE APPLICATION OF ATM TECHNIQUES TO THE LOCAL NETWORK

I Gallagher, J Ballance and J Adams

12.1 INTRODUCTION

Recognizing that the future network will need to provide customers with a very broad and flexible range of services, CCITT Study Group XVIII [1] have recommended that asynchronous transfer mode (ATM) should be the basis of the emerging broadband ISDN standards.

This chapter briefly describes the essential features of ATM, its application to the major service types, and its benefits both to the customer and to the administration.

To gain these benefits, however, requires the provision of broadband access to customers' premises. Hence the introduction of ATM is critically dependent on the introduction of optical fibre into the local loop. The chapter, therefore, goes on to examine the possibilities for fibre access and the support of ATM to provide a high degree of flexibility.

In the local network the need for access to mobile terminals is growing rapidly and may eventually take over a significant section of the fixed terminal market. Here, ATM has potential benefits in two areas: the support of high-speed signalling for call set-up, and the reduction of processing required during hand-off from one radio cell to another. The application of ATM to mobile is discussed in a further section of the chapter.

12.2 THE CASE FOR ATM

When looking to the future it is clear that there is a high degree of uncertainty regarding the requirements of the network. There is uncertainty in the services to be provided, the burden they will place on the network, and the technology to support that burden. To accommodate the uncertainty and be able to respond swiftly to changing circumstances points to the need for a very flexible network.

A feature of such a network would be the ability to introduce any new service without having to provide new types of switching or transmission equipment. Changes to the network tend to be time-consuming and costly requiring detailed foreknowledge of the nature and use of the service. Each addition makes the network increasingly complex and rigid. Thus new services should only need changes to terminal or network processing equipment. This leads to the view that a single, universal network capable of supporting any service would be very attractive. Any service could include a mixture of attributes such as constant or variable bit rate, delay sensitivity or not, and guaranteed delivery.

The support of services with a potentially large range of attributes points to the desirability of connections not associated with a particular fixed bit rate, which means moving away from switching based on the position of a time slot within a frame (position multiplexing as in conventional PCM) to switching based on the routeing information being in a label associated with the time slot (label multiplexing). Other methods of achieving bit rate transparency are possible, such as using only space or frequency division multiplexing; however, it is considered that for the foreseeable future it will be economically favourable to time-divide the use of equipment in addition to these other techniques.

ATM is a label-multiplexing technique which uses short, fixed-length packets, or cells, as the common means to support all services. This is illustrated in Fig. 12.1 where a fixed bit-rate transmission link is divided into time slots of equal length containing ATM cells, each with a label field and an information field. Thus the cells occur synchronously but may be associated with particular connections and services only when needed, in an asynchronous manner.

Thus ATM is seen as a means of implementing a highly flexible network to cope with the uncertainties of the future and not as a more economical way of supporting existing services such as voice or low-speed data. However, it should support them in an affordable manner while opening up new opportunities for the future.

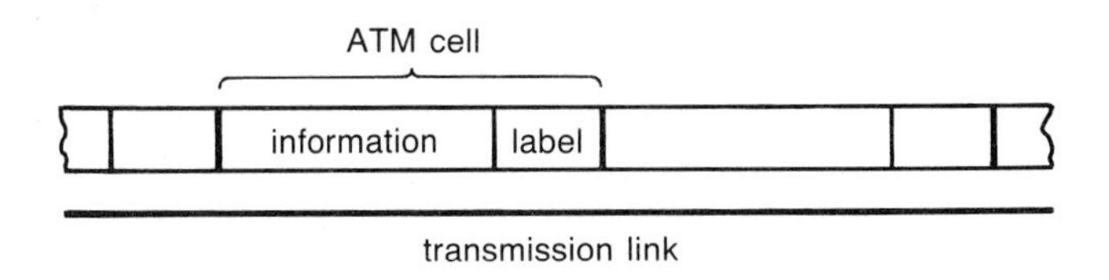

Fig. 12.1 Transmission link divided into ATM links.

Unlike the introduction of digital techniques, where the economies of time-division switching favoured its initial application to the core network, the advantages of ATM are only realized if it is available at the customer's terminal; it is therefore likely that ATM will be introduced first in the access network or customer premises.

12.3 SUPPORT OF SERVICES ON ATM

ATM uses short, fixed-length cells with a minimum of header information. They are short to reduce packetization delay. They are fixed length to make it easier to bound delays through switches and multiplexers. They have a short header to allow the cells to be routed at high speed by means of hardware-implemented routeing tables at each switch.

To enlarge on the claim that ATM is capable of supporting any mixture of services the following paragraphs outline how this may be achieved for the major categories of voice, video and data. In addition some of the features which need to be included in an implementation of an ATM network are discussed.

12.3.1 Voice

The transmission of voice over ATM requires that the signal is digitally encoded and assembled into the information field of an ATM cell. Obviously the longer the cell the longer the delay in filling it. Since voice is likely to be a major user of the network and since it is relatively low bit rate but very delay-sensitive, it is this service which has a major influence on the cell size. For broadband ISDN (BISDN), CCITT have recommended a cell size consisting of a 48 byte information field and a 5 byte header.

12.3.2 Video

For video, ATM could allow any bit rate of transmission to be selected depending on the quality required. Indeed, it may be attractive to develop codecs which have variable bit-rate outputs so that the number of ATM cells transmitted would vary with the rate of change of the picture. In this way it may be possible to take advantage of the picture statistics to gain multiplexing efficiency. However, there remain issues of criteria for call acceptance, quality of service guarantees and charging methods to address before the full advantages of ATM for video can be quantified.

12.3.3 Data

To carry data packets across an ATM network it is proposed to segment them into ATM cells for transmission and switching and then reassemble the original packet at the destination. Thus the network could be used to provide wide area interconnection supporting many different protocols. The high speed of ATM could be used to interconnect LANs based on emerging standards such as FDDI (fibre-distributed data interface, a US 100 Mbit/s optical LAN standard). Currently metropolitan area networks (MAN) are being installed to try out high-speed data services based on early versions of ATM.

12.3.4 Signalling

The current proposal is to base the signalling for ATM on that used for ISDN as defined in CCITT Recommendation I.451 [2]. The signalling messages would be segmented into ATM cells for transmission in a similar manner to data, but it is considered that ATM cells containing certain types of signalling information should be allocated a high priority to ensure their arrival even when the network is congested.

One of the major functions of the signalling would be to set up connections across the network. In ATM these would be virtual connections whereby a unique value is assigned to the label in the cells for a particular connection. The signalling then causes the look-up tables at each switching node to route cells with this label to the correct destination. The look-up tables may also be used to ensure that the switch provides any special handling of the cells, to apply a priority for instance. Thus signalling from the customer to a call-processing entity would request a connection with particular attributes. The processor would then set up the connection and return to the customer's equipment the label value to be appended to each cell.

12.3.5 Implementation

One way of describing the architecture of an ATM network is to use a layered model as in Fig. 12.2. The figure shows three layers — adaptation, ATM and physical-media-dependent (PMD) — used to map different services to different implementations of switching, multiplexing or transmission. The adaptation layer is used to segment the incoming voice, video, data signalling into ATM cells and reassemble the information at the destination. The ATM layer is where all information is in the common form of standard ATM cells. The PMD layer is used to operate on the ATM cells to match them to the particular network implementation, for instance, to insert the cells into an SDH (synchronous digital hierarchy) frame structure, the existing PDH (plesiochronous digital hierarchy), or some other transmission system such as a passive optical network (PON) as described below.

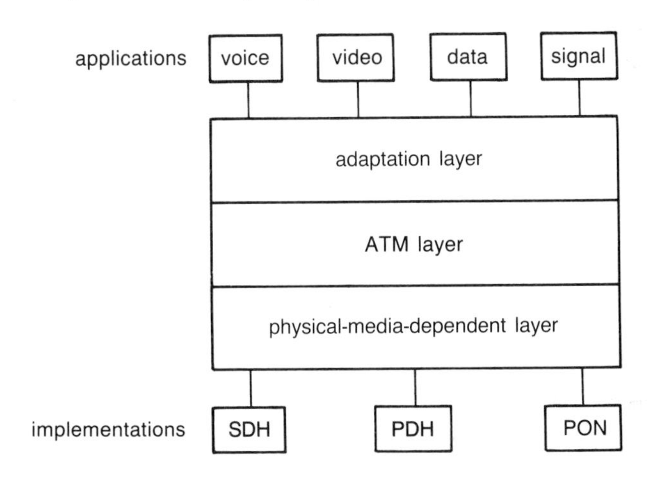

Fig. 12.2 Layer architecture of an ATM network.

12.4 THE BUSINESS SECTOR

The business sector includes those large customers who are heavy users of telecommunications. They have the equivalent of at least 20 telephone lines and may well have private networks including voice switches and data LANs.

12.4.1 Main characteristics

It is the large business sector which could offer the first opportunities to introduce ATM, particularly to satisfy a growing need to interconnect high-speed LANs at rates of up to about 100 Mbit/s. The sector is characterized by many high-occupancy connections already justifying the provision of dedicated optical fibres typically divided into 64 kbit/s circuits. However, such networks are not well suited to high-speed bursty traffic whereas ATM may offer a more appropriate solution.

12.4.2 Access means

It would be possible to build on the installed base of optical equipment to offer ATM at the customer's premises. Instead of the bandwidth of the fibre divided into fixed channels of 64 kbit/s, ATM cells would be used. The customer would therefore have very flexible use of his access link for any mixture of services at any instant. An early manifestation of ATM access has recently appeared in the form of metropolitan area networks (MAN) to provide, initially, a switched multimegabit data service, referred to as SMDS by BellCore [3]. Further development is possible to add the support of other services, such as voice and video, so the MAN could migrate towards an access means for the future BISDN. A parallel path is being followed by some of the private network multiplexer manufacturers in moving to equipment based on ATM in order to multiplex a mixture of services efficiently. In the future this too could become an access means to BISDN.

12.4.2.1 Access structure

Figure 12.3 shows a typical structure for providing ATM access over a MAN to large business customers. Dedicated optical fibre or copper links in the form of a dual bus connect customers to a MAN switching system (MSS). At the customer's end equipment is needed to adapt the services and interfaces to ATM and multiplex the cells on to the bus. At the MSS cells are routed to other customers or to other MSSs according to the contents of the cell header.

Initially the current plesiochronous transmission systems could be used to transport ATM cells on the access link. In future the synchronous digital hierarchy (SDH) with drop and insert techniques could manage the delivery of bandwidth to the customer more flexibly.

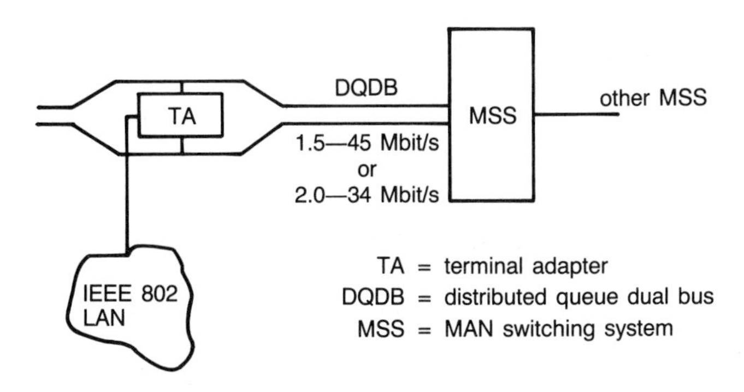

Fig. 12.3 Metropolitan area network.

12.4.2.2 Means

In the current access systems fixed bit-rate circuits are set up and cleared down under administration control according to the wishes of the customer. The allocation of circuits needs to be predetermined by the customer as a semi-permanent arrangement.

It is assumed that the first implementation of ATM access will be in the form of a MAN. In this case a connectionless data service is provided by sending all the ATM cells that make up each data frame to a routeing function that examines the complete frame address and generates the appropriate header to route the cells to their destination.

As connection-oriented services are introduced virtual circuits could be set up by entering the appropriate values in the routeing table at the switch. In the first instance a customer could be allocated a set of labels by the administration to insert into the header of their cells. When the switch received the cells they would be routed according to the value in the label to wherever the customer had requested. Thus the customer could dynamically control the level of traffic and type of service to any of his destinations without the need for action by the administration. As the signalling for BISDN is developed then the allocation and control of virtual circuits will become more and more vested in call-processing software thus increasing the service and connectivity available to the customer.

12.4.3 Exchange equipment

The functions required at a switch based on ATM are to receive cells, route them to the outlet according to the label value in the header, and multiplex them on to the outgoing link. The multiplexing function may need to take into account priorities of cells as it resolves contention for the link. In addition on those links to non-ATM networks adaptation would be needed to convert to and from ATM. A number of techniques are available for implementing an ATM switch based on LAN techniques such as rings or buses or based on interconnection networks [5-9].

In its initial guise as an MSS the switch would be a relatively simple device with cells for a connectionless data service routed on a semi-permanent basis. However, as more services are added and more complex signalling and call processing are introduced, other functions are required. In particular, to give the customer quality of service guarantees for voice, video and data, it may be necessary to provide more complex priority schemes, call acceptance control and policing of sources to ensure they do not exceed their resource allocation.

12.4.4 Customer/terminal equipment

At the customer premises, equipment for adaptation, multiplexing and possibly switching may be required. The initial application of a connectionless data service for LAN interconnect would require the development of equipment to adapt between ATM and, for instance, IEEE 802 LAN standards. Such equipment for MANs based on dual bus structure is now beginning to appear. As adaptation equipment for other services is developed then ATM multiplexers will be needed to resolve contention for the link bandwidth in a way that meets their different requirements.

12.4.5 Evolution strategies

By means of current developments of such MANs it seems possible to introduce ATM into the access to large businesses in order to meet a recognized need for LAN interconnect. Because this is a connectionless service there is no need for highly complex signalling and control software so it is a relatively gentle start on the road towards BISDN. The evolution path can be taken in small steps depending on market requirements.

The data services could be expanded to include the provision of virtual private networks with added services such as databases or protocol conversion. In general, the evolution from MANs to BISDN would allow the trial of a variety of switched video services not easily implemented on existing networks. As the equipment develops, the migration of voice on to ATM may become viable reducing the variety of connections and switches to manage.

By installing ATM islands within large businesses in the form of MANs, and increasing the interconnection between these islands and the services they support, the network could grow gradually and in justifiable stages.

12.5 THE RESIDENTIAL SECTOR

The residential sector includes both domestic customers and small businesses with less than 20 or so telephone lines and limited broadband requirement, i.e. those not warranting a direct fibre link.

12.5.1 Main characteristics

The main characteristic of this sector is that penetration of any one broadband service is likely to be low and, even for telephony, usage rates may be low. Major cost gains may be made if the access network is shared with other services, or if the exchange end equipment for any one service is shared over a number of circuits.

12.5.2 Access means

It is unlikely that a dedicated fibre per customer all the way to the exchange will be cost-effective for this sector; therefore some way of sharing the fibre is desirable. Possible approaches are:

- ring-connecting each customer;
- street-sited multiplexer with short dedicated link to each customer;
- passive optical network (PON) using tree and branch structures.

The ring approach (Fig. 12.4) has problems for growth as it is not easy to add extra nodes (customers) to the ring without either disruption to other customers, or complicated work practices.

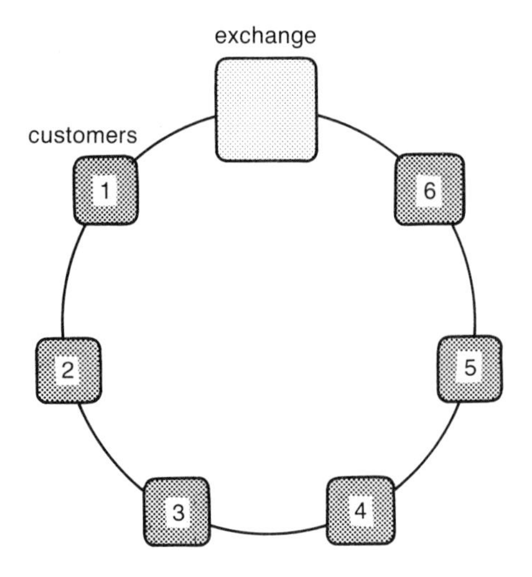

Fig. 12.4 Ring to each customer.

The street-multiplexer approach (Fig. 12.5) requires dedicated street-located electronics. 'Real estate' is required to site these electronics, as well as environmentally acceptable housings. In many situations this may pose serious problems. These street multiplexers may also prove costly to operate and maintain owing to their (harsh) operating environment. Despite these factors, street multiplexers may well be used as a short-term expedient.

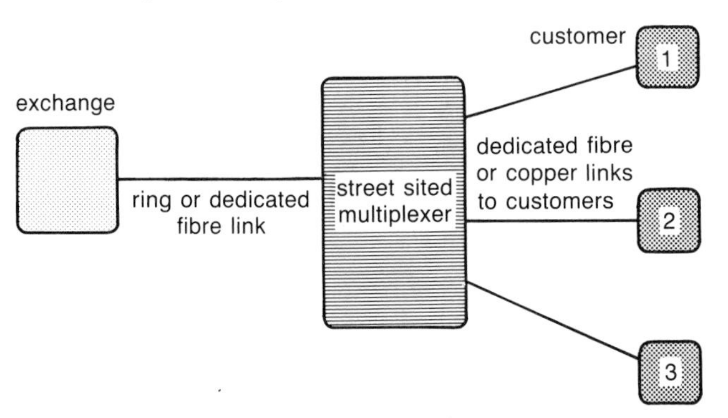

Fig. 12.5 Street multiplexer.

The most practical approach seems to be to use a PON (Fig. 12.6) which may already have been installed for TPON or BPON use (see Chapters 9 and 10).

The link from the customer premises to a street multiplexer is essentially point-to-point and will therefore use techniques similar to those for the business sector. For a PON, however, traffic is shared over the fibre in a point-to-multipoint manner.

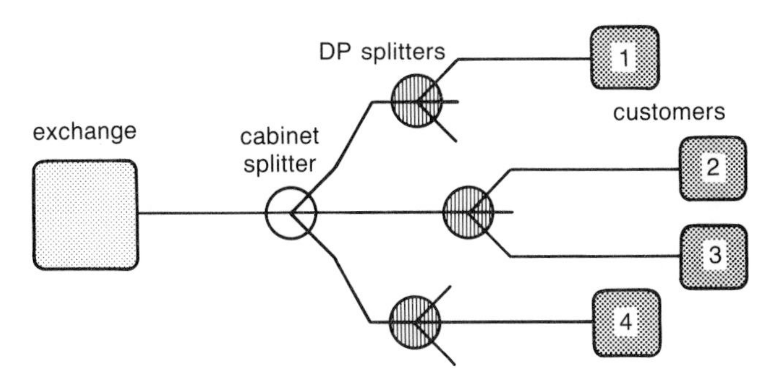

Fig. 12.6 Passive optical network.

Whilst there are undoubtedly several approaches to providing ATM on a passive optical network, the various requirements for serving the residential sector will be illustrated with a description of a potential system called APON (ATM on a passive optical network).

At present it seems unlikely that ATM will become the sole transport method within the residential sector. Indeed it is uncertain whether in this sector ATM will be generally justified on the basis of any one service. Early uses might be for services such as high-speed Prestel to travel agents (probably too small to fall in the business sector), and other as yet unknown services. However, once in, other 'occasional' services such as remote meter reading could use ATM, and it may then be worthwhile to migrate other services to make use of ATM's flexible sharing of bandwidth.

12.5.2.1 Structure

The tree-and-branch structure of a PON maps very well into the existing copper layout with passive splitters located at points equivalent to the conventional cabinet and distribution points.

The fibre network can provide a fairly flat amplitude transmission characteristic from well below 1300 nm to above 1550 nm. In the long term this will permit many wavelengths to be carried, with receivers selecting only those that are wanted. In the medium term, however, technology will only permit a few wavelengths — hence the need to share one wavelength between several customers. As technology options evolve and become cheaper, it will be possible to use a dedicated wavelength per customer, moving eventually to a dedicated wavelength per customer per service. Even when this is possible, it may often be more convenient to time-multiplex several services together.

12.5.2.2 Means

APON takes a stream of ATM cells from the exchange node and broadcasts it to all customer ends. Empty cells will be broadcast when no cells containing traffic are available. Header translation may need to be carried out on the ATM cell headers before they are transmitted. APON also adds a small overhead to each cell which enables each customer end to be polled for its highest-priority cell awaiting return transmission to the head-end and for simple control messages to be sent to each individual customer end.

Each APON customer end will monitor both the polling byte in the APON overhead and the ATM cell header. For each polling byte, the customer end to which the polling byte belongs will transmit a cell (even if it is empty) after the ranging delay, and react to the control information arriving with the polling byte. (The ranging delay compensates for different path delays due to different path lengths.) If the ATM cell header indicates that the cell belongs to this end, it will be extracted and passed to the customer.

12.5.3 Exchange equipment

Shared exchange-end equipment for ATM may be considered to have three main parts — the ATM node, the APON transmission system, and the APON headend controller (Fig. 12.7). It uses its own wavelength over the fibre.

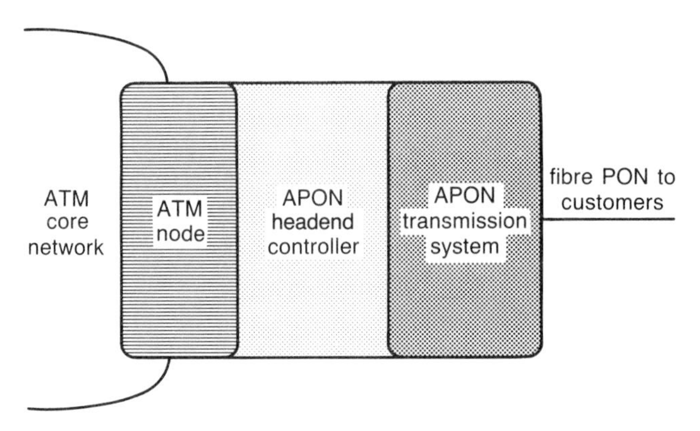

Fig. 12.7 Exchange end.

The ATM node produces a stream of ATM cells from the core network to be distributed via the APON system, and accepts cells from the APON system for onward transmission through the core network. It will either be connected directly to a port on an ATM switch, or at the end of a point to point transmission system connecting the headend to the switch (e.g. an SDH terminal).

The APON transmission system is responsible for taking the stream of ATM cells from the ATM node and broadcasting them to all customers on the APON system. It also schedules return transmissions from these customers to form a non-overlapping interleaved stream of cells back to the ATM node. Any header translation that may be needed is carried out here using data supplied by the APON headend controller.

Whilst it may appear that all customer data is receivable by all customers, proper customer-end design will ensure that only the cells addressed to a particular customer are actually visible to that customer. This proper design will also ensure that a customer cannot easily adapt the equipment to see cells belonging to other customers.

The APON headend controller is mainly responsible for control and maintenance of the APON network. It handles, for example, the initialization and control of each customer end (for range compensation, etc), and the ATM circuit control product for interacting with the ATM node.

12.5.4 Customer/terminal equipment

There are two main functions within the customer end for APON (Fig. 12.8). There is a basic cell receive/transmit function that provides the low-level reception and transmission of cells through APON, and a higher-level controller that provides control, maintenance, and ATM circuit control and policing.

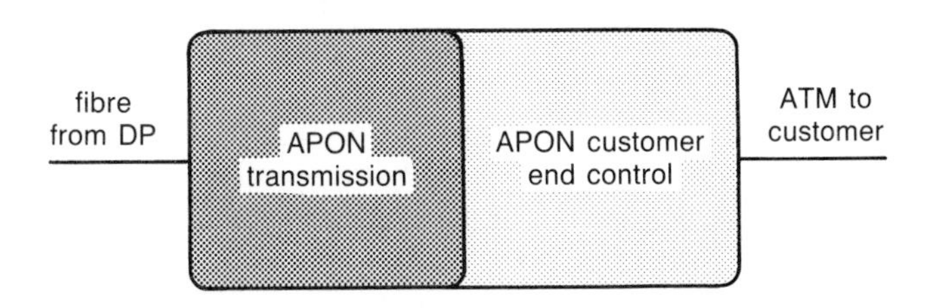

Fig. 12.8 Customer terminal equipment.

12.5.5 Evolution strategies

APON as described here can permit a very effective evolution strategy. Initially it could be installed as an additional wavelength on a TPON fibre infrastructure which would enable the provision of ATM. If a customer grows to need a high-usage ATM link, then it could be provided on a point-to-point basis, either on yet another wavelength on the PON, or on a separate dedicated fibre.

In the longer term sufficient wavelengths are likely to become economically available to allow all traffic for a customer to be multiplexed, on to one wavelength per customer.

12.6 THE MOBILE SECTOR

The mobile sector includes cellular radio, cordless telephones, cordless offices and more limited mobility through personal number services.

12.6.1 Main characteristics

The rapidly growing demand for cellular telephony probably means changes will have to be considered in the system architecture to achieve an orders-of-magnitude capacity improvement in the late 1990s. The use of ATM techniques to distribute and replicate control appropriate for a much higher-capacity system is discussed further in this section.

In the mid-1990s it is expected that some personal number services will be available, including the re-routeing of incoming voice calls to the current customer location. Demand for personal numbers is expected to follow a similar growth pattern to that of cellular radio, and with that growth will come an expansion of services, including broadband, beyond the year 2000.

Demand for cordless PBX equipment is also expected to be heavy following its introduction in the early 1990s. Later, with the introduction of DECT technology, the bit rates available to the customer will be higher and consideration will be given to methods of adding cordless services to dumb ATM dual bus/star customer networks. The 'dumbness' comes from the early targeting of popular broadband services, i.e. simple virtual path cross-connection and connectionless data services. The technique of using 'intelligent' ATM packetizers to support mobility which can be added on to dumb networks is described further in this section.

12.6.2 Access means

Access for personal number broadband services could be via the ATM local exchange, initially to one or more special switching centres containing location information. The access means is similar to that described in the previous sections.

Access for today's cellular radio is based on analogue technology. However, the early 1990s will see the introduction in Europe of digital cellular radio, namely the CEPT groupe speciale mobile (GSM) pan-European network in which TDMA techniques will be used to provide a number of channels per carrier. However, the greater the number of channels the higher the peak power required from the hand-portable terminal's RF amplifiers and batteries; a compromise value of eight channels per carrier has been selected for GSM.

Greater capacity can be achieved through the use of radio microcells in which one base station would typically cover an area of 100 m radius. This requires an increase in the number of base stations of about 100-to-1 compared with the normal radio cell size. It is expected that the use of microcells will be localized to centres of high traffic demand in urban areas.

There seems to be little advantage in using ATM to support access between the handset and the base station. If developed at all it would be targeted at the cordless office environment to make the greatest use of the higher bandwidth available. However, the introduction of the DECT cordless office terminal around 1994 is expected to include the capability for supporting bandwidths of up to approximately 300 kbit/s. This may remove the need for such development work, at least in the medium term.

After the signal from the terminal has reached the base station it can be packetized into ATM cells. Depending on what information is stored in the header of these ATM cells they may be routed to different destinations at different times as required using only a dumb ATM network.

12.6.2.1 Structure

The connection of a number of base stations to an ATM switching node may be achieved using a passive optical network tree-and-branch structure, although other structures are not precluded, e.g. a ring. If each PON is assigned a single address for destination-routed ATM cells then, on arrival at the destination PON, such cells will be broadcast to all base stations attached to the PON. The further identification of which ATM cells are associated with a given base station comes from the connection number stored in a separate 16-bit field in each cell (see Fig. 12.9).

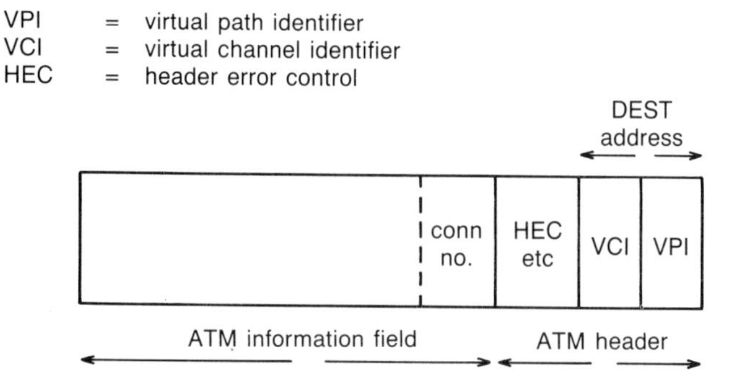

Fig. 12.9 ATM packet structure.

The structure of the destination address field consists of two parts as shown in Fig. 12.9, i.e. the virtual path identifier (VPI) and the virtual channel identifier (VCI). The VPI partitions any ATM cell stream into a number of separate payloads [10]. It may be used to distinguish between payloads with differing quality-of-service requirements (i.e. different tolerances to cell loss rates and end-to-end delay) and has a limited capability to distinguish between payloads requiring separate routes.

To obtain destination addressing of hundreds of thousands of PONs located over the whole of the UK, as may eventually be required beyond the year 2000, it is proposed that a few VPI values are reserved. Each VPI value represents one address 'block' of 64k addresses as specified by the value in the 16-bit VCI field — therefore perhaps eight values need to be reserved. The total address space, which is an integer multiple of 64k, has been called an ATM zone [11].

Half of the reserved VPI values are used to carry signalling information, and such cells will also carry a higher priority through the ATM network (i.e. such cells will be less likely to be discarded). This defines a signalling path in both directions between the two current destinations of the calling and called customers of any given connection. This signalling channel is a high-capacity (broadband bit rate) channel capable of transferring information quickly between the two ends (e.g. for a 1000 km route through the ATM network, the mean one-way delay including packetisation is about 10 ms). When not in use, the channel capacity can be temporarily assigned to other users according to VPI multiplexing principles such as in Adams [10]. Additional signalling paths exist between a base station and controllers elsewhere in the network, the paths being identified by suitable signalling VPI values plus particular VCI values. The remaining reserved VPI values are used to carry voice cells base-station-to-base-station with no change to the cell header.

PON destination addresses operate as group addresses, i.e. the address assigned to each PON is also valid for all other PONs which have adjacent radio cells. At the ATM cross-connect (see Fig. 12.10) ATM cells carrying a group address are automatically broadcast to all the relevant PONs.

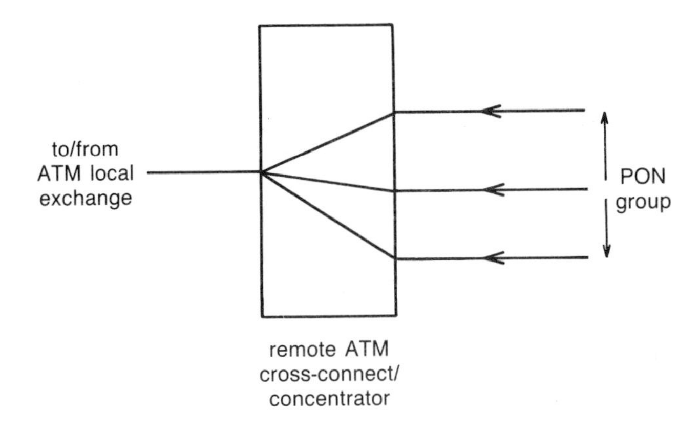

Fig. 12.10 Signalling path for PON group.

12.6.2.2 Means

The primary purpose of the structure described above is to achieve base-station-to-base-station control of tracking. This removes processing load from a central controller and hence removes a potential processing bottleneck particularly as the system evolves towards the greater use of microcells. As either customer moves the ATM zone automatically provides a new voice path and a new path for the transfer of signalling information. There is minimal delay and minimal processing load involved in setting up a different path since the route to any given destination is already predetermined. The capacity previously used on the 'old' path and no longer required is automatically made available to other users according to VPI multiplexing principles [10]. The technique also enables intelligent tracking to be added to a dumb network.

For each mobile terminal currently associated with a particular radio cell/microcell, all of the base stations within its group address will store that terminal's current PON address, current destination PON address and connection number, together with the 'anchor' controller address (explained below). This information is loaded into tables at each base station by a broadcast message from the opposite end, i.e. each end of a call in progress is responsible for updating the group tables of the other end.

The mechanism for updating group tables is triggered whenever a terminal moves to an adjacent radio cell, minicell or microcell. In this case the new base station will send a signalling message to the address of the destination PON, containing the customer identity at both ends, the 'anchor' controller address, the destination PON address and information on the updated near-end PON address and connection number, if the values of these parameters have changed. In response the base station associated with the terminal at the other end will send its current values of the same list of parameters regardless of whether there have been any changes or not.

Connection numbers need to be changed whenever there is a change in the PON address. Connection number blocks containing free connection numbers can be pre-assigned to each base station in a PON group so that the selection of a new connection number can be made autonomously by any base station as the terminal moves from PON group to PON group.

ATM packetizers at the base stations automatically load the destination PON address into the headers of voice cells. This is the only action required to route the cell. Cell counters are used at the base stations for the purposes of charging. On moving to a new base station, the 'old' base station sends charging information using ATM signalling cells with a destination address set to the 'anchor' controller. This controller is the only fixed point in the architecture — hence the name 'anchor'. The 'anchor' controller becomes associated with a given call at the time of call set-up. Its identity then remains fixed throughout the call. Using the zone concept an ATM route always exists to any 'anchor' controller from any base station.

12.6.3 Exchange equipment

ATM cross-connects and ATM exchange equipment must be configured for zonal operation. This means that the VPI/VCI address is translated to the same value going from the inlet to outlet of the switch and hence is preserved over the route from end to end. At the switch inlet the VPI and VCI numbers are used to reference special routeing tables for zonal cells and an internal routeing label is created. This label has sufficient room to store the destination address as well as containing information to route the ATM cell to the correct outlet.

The routeing tables can take account of fault conditions by storing 'first preferred' and 'second preferred' outlets together with the appropriate routeing tags for internal switch operation. An automatic fault alarm would cause the routeing tables to switch to the second-preferred outlet if appropriate.

The routeing tables are the only resource which is special to the mobile users and the same ATM switches can be used for other BISDN services.

12.6.4 Customer/terminal equipment

The main functions at the customer end are the ATM packetizers and the end-to-end tracking control both of which are located in the base stations. A higher-level controller is also required to provide for charging, maintenance, some circuit control functions and policing. Terminal equipment compatible with the GSM system may be used.

12.6.5 Evolution strategies

The first phase may be to offer mobile services over dumb ATM dual bus networks, with intelligent packetizers in the customer's network and at the ATM local exchange to reflect the correct control information back to the customer. The call is then routed through the ATM network to the corresponding local exchange of the called customer, which may be either fixed or mobile.

This arrangement is targeted at the cordless office but is capable of evolving further, either to support an increased-capacity voice cellular radio network (with greater replication of control) or higher bit-rate services.

12.7 THE WAY FORWARD

ATM is an attractive solution which can flexibly meet the uncertain transmission and service demands of the evolving future and eventually provide a universal, multiservice network. However, it is not a technique for saving costs on existing services and needs to be justified on its flexibility and potential for easing the introduction of new services.

The earliest opportunity appears to be the support of high-speed LAN interconnection which could justify the introduction of ATM in the form of MANs or proposed systems such as PONs. The first step would not be a bold one and the rate of evolution could be tuned to the market requirements.

On a single ATM 'island', in addition to data, new viewphone, videoconferencing, or interactive video services could be demonstrated. The advantages of high-capacity signalling and the self-routeing nature of ATM cells suggest mobile applications as another area for an early demonstrator. The applications range from cellular through cordless phones to customers moving from one fixed terminal to another. Customers moving or expanding their offices and creating a 'churn' in access provision could also be treated as mobile.

Having begun with ATM 'islands' the degree of interconnection could then be increased to form an ATM network covering the major business centres.

Around the world considerable research effort is being devoted to the study of ATM, particularly in Japan, the USA and in Europe through the collaborative RACE (research into advanced communications for Europe) programme. The aim is to develop the standards during the current CCITT plenary which ends in 1992 with the first trials soon after, and the first equipment introduced by 1995.

REFERENCES

1. CCITT Recommendation I.121.

2. CCITT Recommendation I.451.

3. Bellcore Technical Advisory TA-TSY-000772 'Generic system requirements in support of switched multi-megabit data service'.

4. Adams J L and Falconer R M: 'Orwell: a protocol for an integrated services local network', BT Technol J, 3, No 4, pp 27-35 (October 1985).

5. Final report of the COST 202 bis experts group on ATM switch structure (November 1988).

6. Yeu Y S, Hluchyj M G and Acampora A S: 'The knockout switch. A simple, modular architecture for high performance packet switching', AT&T Bell Laboratories, Holmdel, New Jersey 07733.

7. Bernabei F, Forcina A and Listanti M: 'On nonblocking properties of parallel delta networks', INFOCOM'88, New Orleans, USA, pp 326-333 (1988).

8. Giacopelli J and Hickey J: 'Applications of self-routing switches to LATA fibre optic networks', Bell Communications Research.

9. Huang A and Knauer S: 'STARLITE: A wideband digital switch', AT&T Bell Laboratories.

10. Adams J L: 'The virtual path identifier and its applications for routeing and priority of connectionless and connection-oriented services', International Journal of Digital and Analog Cabled Systems, 1, No 4, pp 257-262 (1989).

11. Foster G and Adams J L: 'The ATM zone concept', Globecom '88, Florida, pp 672-674 (November 1988).

13

DIGITAL SIGNALLING IN THE LOCAL LOOP

R A Boulter

13.1 INTRODUCTION

Signalling in the local loop has always been digital in nature. The basic telephone uses signalling at 10 pulses per second, either by using breaks in the d.c. loop of the pairs between the telephone and local exchange, or by sending pulses of multifrequency tone. Similar techniques have also been used between PABXs and the local exchange.

The advent of the integrated services digital (ISDN) network has seen the need for more powerful and faster digital signalling techniques which provide not only the terminal with a more flexible and powerful signalling capability but also provide the same capability in the direction from the network back to the terminal, something unknown in the slower decadic signalling systems. In addition a requirement has also arisen in ISDNs for a signalling capability not only during the set-up and clear-down phases of the call but also during the call.

The first high-speed digital signalling system with these advanced features, was introduced in the UK for BT's Pilot ISDN (or IDA services) and was called digital access signalling system no 1 (DASS1) [1]. This signalling system has now been enhanced into DASS2 [2] and a derivative, DPNSS (digital private network signalling system) [3], used in digital private networks. CCITT Recommendations on local access signalling are now contained within its I series recommendations and it is these that now form the basis of BT's ISDN and it is expected that these will gradually replace the proprietary BT signalling systems.

The introduction of optical fibre, and as a consequence digital techniques, for carrying telephony in the local loop is now giving a further impetus to the use of digital signalling and is likely in the future to be more dominant than ISDN.

13.2 THE INTRODUCTION OF DIGITAL TECHNIQUES

Until fairly recently the whole of the public switched telephone network (PSTN) was based upon analogue transmission and switching techniques. The telephone converted the acoustic waves of the speaker into an electrical signal occupying a bandwidth of the order of 4 kHz (300-3400 Hz). This was transmitted over the telephone network in this form to a remote telephone which then reconverted the electrical signals back to acoustic waves. Between the telephone and local exchange, in what is known as the local network, the go and return paths were provided over the same twisted pair, whilst in the main network four-wire techniques were often used.

In the same way that the speech signal was carried in a 4 kHz analogue form, signalling was also constrained to the same frequency band and always constrained to the same channel. In the local network, loop disconnect signalling was and is still used in the majority of cases between the telephone and the local exchange. In Strowger exchanges the equipment is operated directly by the breaks in the d.c. current and this same technique was used over the junction network by the local exchange to operate the main switching exchange. When a carrier system was encountered, this signal had to be translated from the near d.c. loop disconnect signal to a voice frequency signal within the 300-3400 Hz channel band (generally 1800 Hz). Although this signalling by its very nature is digital, it is low-frequency and very restricted in nature. The advent of the adoption of digital techniques in public telephone networks was to herald a new step in the features and facilities that could be offered by a high-speed digital signalling system.

The introduction of digital techniques in the network started with digital transmission. Although digital transmission, in the form of telegraphy, of course pre-dates analogue telephony by a number of years, the carrying of the analogue telephony signal in digital form was not invented until the 1930s when pulse code modulation techniques [4] were first patented. However, this technique, together with time division multiplex techniques, did not become commercially viable until the 1960s when they were first used in the junction network for routes of over 8 miles. The original systems in the UK carried 24 channels in a 1536 kbit/s structure but now these have been replaced in Europe by 2048 kbit/s systems carrying 30 channels in which a separate time-slot 16 is reserved for signalling [5]. (In the USA a 24-channel system based on 1544 kbit/s has been adopted.)

The first modern form of digital signalling was therefore used in these original 2048 kbit/s PCM transmission systems with each channel in the system being allocated four bits in time-slot 16 which were then used to convey the loop condition and other states. Specific signalling bits are therefore still associated directly with a particular channel and at the multiplexer they are translated back to an in-band signalling system for the exchange to handle.

Specialized data networks, however, have seen the introduction of message-based signalling systems, for example X.25 in the packet-switched network, but this type of signalling was not introduced into the PSTN until the introduction of stored program control (SPC) exchanges. Common-channel message-based signalling systems, where messages relating to different connections are statistically interleaved on a common channel, were then introduced to carry messages between exchanges. In the UK this occurred at the same time as the introduction of digital switches in the form of System X but in other countries they were used on analogue SPC exchanges. Loop-disconnect and multifrequency signalling, however, continued to be used between the customer's terminal and the local exchange.

In the same way that technology made the use of digital techniques in the transmission field more attractive, so digital exchanges became cheaper to implement and maintain. These digital exchanges are designed to handle 64 kbit/s channels transparently using both space- and time-division digital switching. Main network digital switches became particularly attractive because of the existence of digital transmission and as a consequence the saving that can be made in analogue to digital conversion equipment. These digital main network exchanges together with the digital transmission form what is known as the integrated digital network (IDN). The first System X digital exchanges in the UK were put into the main network in 1980 and now all the main network exchanges in the UK are digital and fully interconnected by digital transmission systems.

Digital local exchanges also became attractive, avoiding the analogue-to-digital conversion on the main network side of the exchange, and these are currently being introduced into the network. The conversion from the analogue signal produced by the standard telephone to a digital signal occupying 64 kbit/s is now occurring at the input of the local exchange. Therefore connections across the main network between local exchanges are able to provide:

- a completely transparent 64 kbit/s channel;
- a powerful digital signalling capability;
- the flexibility provided by stored program control exchanges.

These facilities have been provided in order to support telephony in the most economical way possible. They have, however, set the scene for the development of what has become known as the ISDN (integrated services digital network) and the economic introduction of digital signalling in the local loop.

13.3 BT'S DIGITAL ACCESS SIGNALLING

13.3.1 BT's pilot ISDN signalling

CCITT recommendations state that 'ISDNs will be based on the concepts developed for telephony IDNs and may evolve by progressively incorporating additional functions and network features including those of any other dedicated networks such as circuit switching and packet switching for data so as to provide for existing and new services'.

In common with most telecommunications administrations BT wished to exploit fully the features of its emerging integrated digital network (IDN) in order to extend the range of services offered to its customers. BT saw the early introduction of an ISDN pilot service as a means of gaining practical experience of developing and operating an ISDN and stimulating customer interest, and planning for this was started in 1979.

At this time it was decided to offer two new types of customer access and to market them under the name integrated digital access (IDA). These were a single-line access and a multiline access as shown in Fig. 13.1 and were structured as follows:

- single-line IDA — a 64 kbit/s channel for speech or data
  — an 8 kbit/s channel for data only
  — an 8 kbit/s channel for signalling only

  total 80 kbit/s plus additional framing information;

- multiline IDA — 30 × 64 kbit/s channels for speech or data
  — 1 × 64 kbit/s channel for signalling
  — 1 × 64 kbit/s channel for framing and alarms

  total 2048 kbit/s as for the PCM primary multiplex structure.

In both accesses a channel, separate from the main user channels, has been identified specifically for signalling and in order to introduce these new services a message-based digital customer-network signalling system needed defining. This new signalling system must provide the speed and repertoire appropriate to the full range of services and facilities which can be provided by an ISDN. At the time when BT was preparing its specification for the pilot service no CCITT recommendations on digital access structures or signalling systems were available and so a totally new signalling system, DASS (digital access signalling system) no 1, was defined.

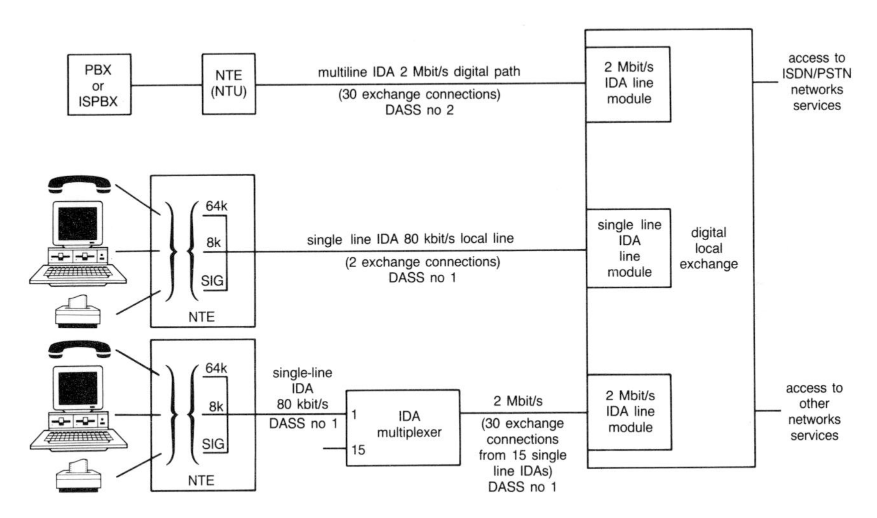

Fig. 13.1 IDA access arrangements.

DASS1, as defined for the pilot ISDN, will initially only support basic services and the same supplementary services as defined for telephony customers and normally accessed by those customers using MF4 signalling (such as call division, abbreviated address, three-party call, etc) along with two new services for data customers (closed user groups, and called and calling line identities). This was in order to minimize exchange software developments required for the pilot ISDN. There is also a need for users to indicate at the beginning the type of call — voice or data.

The time division multiplexed (TDM) channel structure of the ISDN customer interface allows the use of common channel signalling and for it to be related to the traffic channel in a logical rather than a physical sense. Thus DASS was originally able to support the multiline access service as well as being used to provide access over a 2048 kbit/s link to a remote 2048 kbit/s multiplexer supporting 15 single-line IDA customers.

The proposed introduction of DASS1 for multiline access to PABXs, together with the development of digital PABXs and digital leased circuits, led to the need for a digital inter-PABX signalling system. BT and a number of UK PABX manufacturers collaborated on the definition of a signalling system based upon DASS1 but enhanced to meet the inter-PABX signalling requirement. This further signalling system was called digital private network signalling system (DPNSS). During the definition of DPNSS it became apparent that it was desirable to align more closely certain of the messages of DASS with those of DPNSS. At the same time proposals were being made to enhance the repertoire of signalling messages in order to provide more facilities to the user. An enhanced version of DASS1 was therefore defined, called DASS2, which would enable PABXs to have a common type of signalling for both inter-PABX links of private circuits and PABXs to network signalling. The common features of DPNSS and DASS enable them to be interleaved on a common signalling link such as time-slot 16 in a 2048 kbit/s multiplex structure.

The structure of DASS and DPNSS is based upon the first three layers of the ISO (International Standards Organization) model for OSI (open systems interconnection) [6]. The first three layers are defined as:

- layer 1: physical — 8 kbit/s and 64 kbit/s signalling channels of the single and multiline access,
- layer 2: link access protocol — based upon high-level data link control (HDLC) procedures,
- layer 3: network — a new defined set of messages for call control and maintenance,

and are described in more detail in the following sections. At layers 1 and 2 DPNSS is identical to DASS2 except for minor changes necessary to introduce symmetry and to provide virtual signalling channels. At layer 3 both systems use similar messages and coding structures but DPNSS has additional features to support the particular needs of private networks.

13.3.2 Link access protocol (layer 2)

The link access protocol (LAP) is used to provide a secure method of data transfer ensuring reliable exchange of call-control messages between the user and the network. The LAP incorporates features to enable the detection and recovery from transmission errors. Furthermore protocol elements are included to allow the following functions:

- identification of the traffic channel to which a given signalling message related (this is done by assigning a separate LAP to each traffic channel);

- interleaving of signalling messages associated with different traffic channels;
- message sequencing.

The DASS2 LAP is based on the use of unnumbered information (UI) frames as defined in the ISO high-level data link control standard (HDLC). The HDLC format at layer 2 as shown in Fig. 13.2 has the following fields.

flag 01111110	address	control	variable length information field	frame check sequence	flag 01111110
8 bits	8 bits	8 bits	multiple of 8 bits	16 bits	8 bits

Fig. 13.2 Layer-frame format.

- The flag acts as a unique delimiter, and additional circuitry ensures that the flat pattern never occurs in the bits in the rest of the frame.
- The address field is used to indicate whether the message is a command or a response, a signalling or maintenance message and to which channel it refers.
- The control field is used to control the interchange and acknowledgement of layer 2 frames.
- The information field contains the signalling or layer 3 message.
- The cyclic redundancy check (CRC) bits enable the receiver to determine whether any errors have been incurred during transmission. Complete error protection is obtained by repeatedly retransmitting the frames until a positive acknowledgement is received from the receiver (this is referred to as a 'compelled signalling system').

13.3.3 Layer 3 — message structure and sequences

At layer 3 a complete set of call-control messages have been defined to enable the establishment and clearing of connections and supplementary services. In addition to these, messages have been defined for monitoring and maintenance functions. Each message contains an integral number of octets up to a maximum of 45. The longest of these messages is generally the initial service request while the call-accepted message is only a single octet. The first octet contains the message name (type) while subsequent octets contain a variety of information elements. As in the call-accepted message this first octet is all that is needed to identify an event in the call sequence.

Other messages contain a number of mandatory information elements (e.g. the message used to request call establishment always contains a service indicator code describing the characteristics of the required call). In addition some messages contain optional information elements such as network supplementary service requests, which are coded in a flexible manner to allow for future extension.

A PBX or terminal requests call establishment by sending an initial service request message (ISRM). This message contains a service indicator code (SIC) and the called address. The SIC contains an indication of whether an end-to-end digital path with common channel signalling is essential for the call and also contains a description of the characteristics of the calling terminal (i.e. voice or non-voice including data transmission characteristics). This information is eventually passed on to the called customer so that compatibility between the calling and called terminals can be confirmed before the call is accepted. The sequence of messages for the establishment of a simple call (i.e. without any requests for supplementary services) following on from this request is shown in Fig. 13.3.

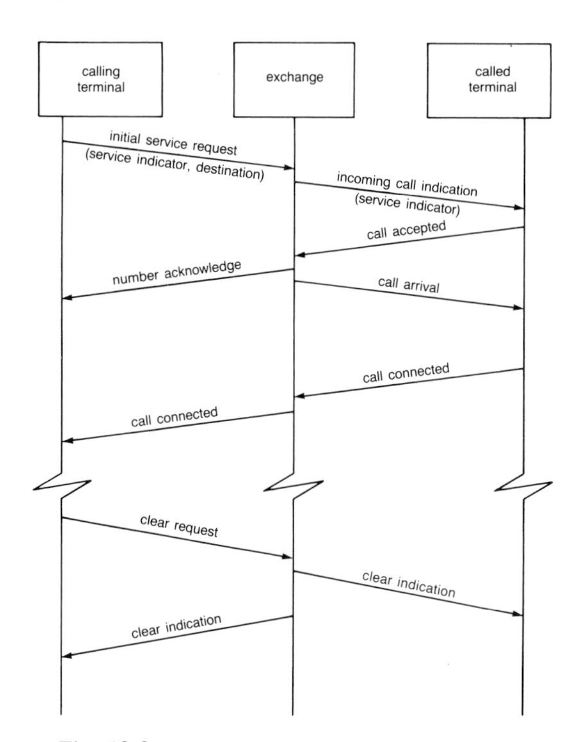

Fig. 13.3 Layer 3 basic call message sequences.

In the early transition phase from an analogue to digital network, complete digital routeing over the network could not be guaranteed; therefore two categories of ISDN call were defined. Users indicate the category of call required within the SIC of an initial service request message as previously described. The category 1 call indicates that a digital routeing with common channel signalling between calling and called subscribers is essential. In general this means that some form of data is to be carried. A category 2 call is one in which a digital path with common channel signalling between calling and called subscribers is not essential. In general this means only a telephony call is required. If a category 2 call is established with an all-digital routeing, all the ISDN supplementary services are available as for category 1 calls. However, if a digital routeing is not available the call will still be established but the only ISDN supplementary service supported will be call charge indication. Furthermore, the calling party will be informed by the network that the call is not suitable for data communications.

In addition to the ISDN call types, a third type of call known as 'telephony' may be established by callers on the analogue network. Telephony calls may only be used to support voice communication.

A summary of DASS2 services available with the two ISDN call types is given in Table 13.1.

Table 13.1 Types of service supported by DASS 2.

Basic call types	*Supplementary services*
Cat 1 — digital path required service characteristics may be voice or data — terminal rate specified for data calls	network address extension user-to-user signalling closed user group calling/called line identity call charge indication
Cat 2 — digital path preferred — service characteristics are voice — continues as telephony call if transparent 64 kbit/s routeing is not possible	as for Cat 1 if fully digital path provided

13.4 SIGNALLING TO REMOTE MULTIPLEXERS

In the digital environment it became possible and cost-effective to locate certain of the functions of the main local exchange remotely from the main processor and route switch. Initially small local exchanges were replaced by a remote concentrator unit dependent upon a larger local exchange. In this instance BT's inter-exchange signalling system based upon CCITT No 7 signalling system was used to connect the concentrator to the local exchange.

A further stage of equipment deployment is now becoming more attractive, especially with the use of optical fibres in the local network; this is to locate the first 2048 kbit/s multiplexing stage of the concentrator remotely.

This first occurred in BT's pilot ISDN, when a multiplexer connected to an early System X exchange with ISDN capability was located in analogue exchange; then DASS1 was used to support 15 single-line customers. However, a more long-term use of a multiplexer is seen with the introduction of optical fibre cables in the local network, initially for business customers and later for residential customers.

To meet the demands of business customers for rapid service provision and circuit flexibility, BT has begun to introduce optical fibre links between the local exchange and these customer's premises. The concept is shown in Fig. 13.4 and from this it can be seen that a 2048 kbit/s multiplexer, called the customer service module (CSM), is located in the business premises to carry private circuits, telephony and later ISDN traffic over the local loop back to a service access switch in the local exchange. Standard 2048 kbit/s structures are used and it is planned to use DASS2 for the switched traffic. However, the multiplexing of analogue telephony circuits has placed additional requirements on the DASS signalling system which has required the definition of additional messages in the layer 3 message set.

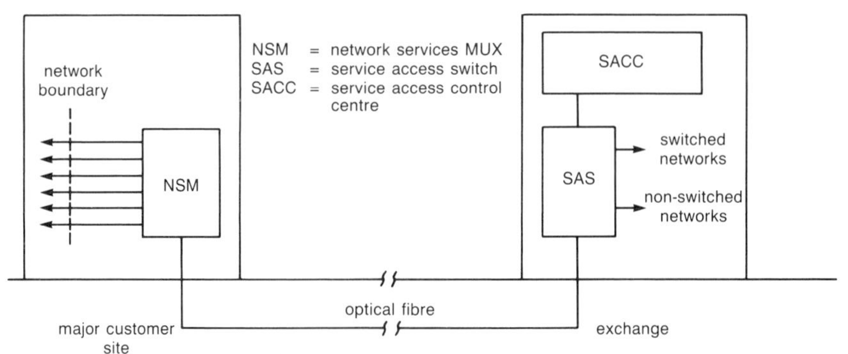

Fig. 13.4 Flexible access system concept.

As well as providing optical fibre to large business customers new techniques are being investigated to introduce optical fibres into the network to support residential customers. These include the siting of remote multiplexers, supporting 30 customers, underground with short twisted-pair copper feeds into the customer's premises and a dedicated optical fibre feed back to the exchange. Passive optical splitters and couplers are also being used in a field trial; these combine the signals from several multiplexers and eventually replace the multiplexer itself when the fibres are taken directly into the customer's premises (see Chapter 9). All these new techniques and networks require digital signalling. Initially they will be based on BT's enhanced DASS2, but it is expected that in the future international recommendations will emerge from CCITT for these architectures.

13.5 CCITT ACCESS SIGNALLING RECOMMENDATIONS

At quite an early stage in the CCITT studies it became apparent that the emerging standards for signalling in an ISDN would be significantly different from BT's initial ISDN access arrangements. The notable differences arose from the CCITT requirement to support two 64 kbit/s B channels and a 16 kbit/s D channel for signalling and for the interface to support a passive bus arrangement which enabled up to eight terminals to compete for access to either B-channel.

BT were anxious to gain experience of the CCITT standard interface at the earliest opportunity. Consequently, following the publication of the CCITT Red Book in 1984, BT began to specify an ISDN basic rate access arrangement using the I-series interface described by Recommendation I.420. BT placed a contract for the development of a multiplexer and a network termination which supported the I.420 interface. Initially this was based on the 1984 Red Book Recommendations; however, with the co-operation of the manufacturer, the interface specification was kept in alignment with changes made to the recommendations and in 1990 supported the test and development phase of BT's ISDN-2 service. This service provided an ISDN basic access interface in line with current international recommendations and was launched as a full national service in February 1991.

The reason for developing a multiplexer, rather than support customers directly on the local exchange, was to avoid the long lead times experienced on exchange developments and to enable more companies to compete for the contract. The multiplexer therefore had to interface to the exchange via an existing interface and the 2048 kbit/s interface supporting DASS2 was the obvious choice. The multiplexer was therefore required to convert the protocols of the I.420 interface to those of DASS2 as well as provide a transmission system capable of supporting 144 kbit/s over the local loop. During 1990 and 1991, 6000 multiplexers were deployed in the network to support BT's ISDN-2 service.

The latest version of the I Series Recommendations is available in the CCITT 1988 Blue Book. The signalling recommendations relate to procedures to be followed across the interfaces at the 'T' and 'S' reference points. These reference points are identified in the ISDN reference model are shown in Fig. 13.5 and, as can be seen, they are buffered from the local loop by the network termination (NT1). However, the NT1 is transparent to the layer 2 and 3 procedures of the protocol which therefore pass transparently across the local loop to the local exchange.

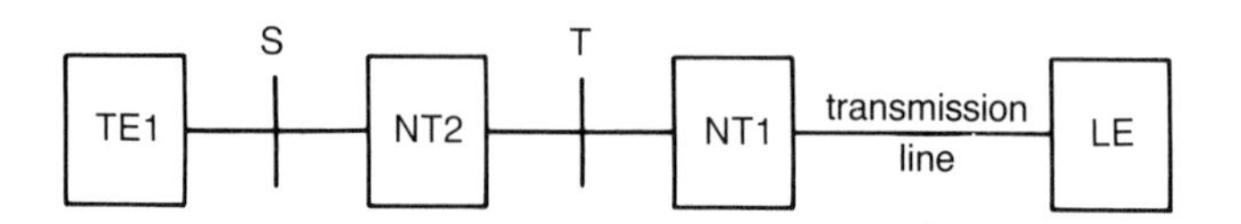

Fig. 13.5 Reference configuration for the ISDN user-network interfaces.

At the 'T' reference point two interfaces are defined, one for basic access at 144 kbit/s and one for multiline access at 2048 kbit/s, and these are contained in recommendations I.420 and I.421 respectively. These recommendations then call up I.430 and I.431 respectively for the layer 1 descriptions, and common recommendations for layer 2 in I.440/1 and layer 3 in I.450/1. The relationship between these recommendations is illustrated in Fig. 13.6. The characteristics of the signalling procedures will now be described in more detail and for completeness mention will be made of the layer 1 features of these recommendations for basic rate access.

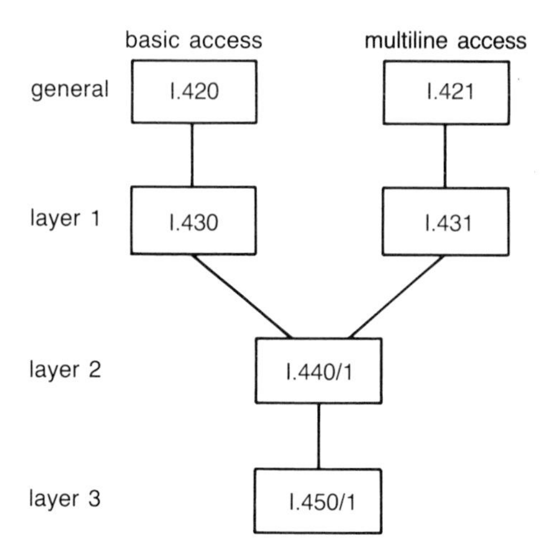

Fig. 13.6 Relationship between CCITT interface recommendations.

13.5.1 Layer 1 — basic access

In the UK the public network terminates in the network termination 1 (NT1). The socket in the NT1 is the regulatory boundary between the public network and the liberalized area of customer premises equipment. It therefore provides the layer 1 interface between the local loop transmission system and the layer 1 interface. Into this socket is plugged the four-wire bus often known as the 'S' or 'T' bus because it is at the 'S' and 'T' reference points in the CCITT reference model. This bus can operate in two modes, point-to-point or point-to-multipoint. In the point-to-point mode one terminal equipment (TE) is connected at the end of up to about 1 km of cable. In the point-to-multipoint mode up to eight terminals can be connected in parallel anywhere along the bus, but the bus length is now limited to less than 200 m.

Over this bus passes the two B channels, which are transparent 64 kbit/s user channels, as well as the D channel, which is the 16 kbit/s signalling channel, and other bits used for miscellaneous purposes such as frame synchronization, bringing the gross bit rate on the bus to 192 kbit/s. The B channels contain the user data which is switched by the network to provide an end-to-end transmission service. A B channel path is established by signalling messages in the D channel. In a multi-terminal situation all terminals have access to the D channel by the use of an access procedure but each B channel is allocated to a particular terminal during call set-up and cannot be shared between terminals.

Power feeding is also provided across the interface. The direction of this power transfer depends on the application, but in BT's network power will be provided from the network to the terminal in order, for example, to maintain a basic telephone service in the event of local power failure. However, to save power fed from the local exchange, both in the local network and the NT1, an activation/deactivation procedure is also defined.

Finally, a connect/disconnect indication is used by layer 2 to determine whether a terminal is plugged into the bus. This is necessary as each terminal is given a random, unique identity (its TEI value) by the network. If disconnected the terminal must forget its original identity to prevent duplication when reconnected. The connect/disconnect indication is achieved by monitoring the presence of d.c. power on the bus.

The layer 1 structure for supporting these functions is shown in Fig. 13.7. The frame is 48 bits long and lasts 250 μs resulting in a bit rate of 192 kbit/s with each bit approximately 5.2 μs long. Figure 13.7 shows that there is a 2-bit offset between transmit and receive frames. This is the delay between a frame start at the receiver of a terminal and the frame start of the transmitted signal. It also shows a 10-bit offset between the D channel leaving a terminal, travelling to the NT1 and being echoed back in the E channel. This 10-bit delay is made up of bus and transmission delays in the NT. A frame contains several L bits; these are balanced bits to prevent a build up of d.c. on the line.

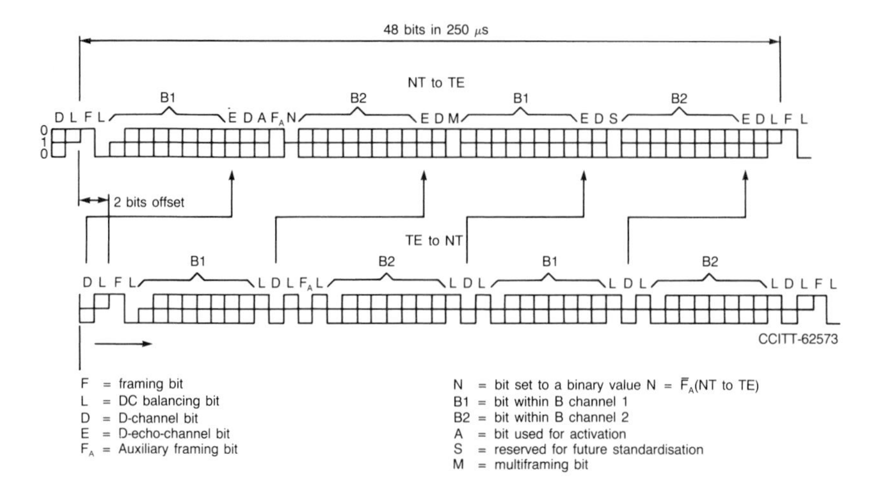

Fig. 13.7 Basic access frame structure.

13.5.2 Layer 2

The layer 2 structure is very similar to that described for DASS and shown in Fig. 13.2, but differences exist in the format of the address and control fields. The control field is one or two octets depending on the frame type and carries information that identifies the frame and the layer 2 sequence numbers used for link control.

The address field enables layer 2 multiplexing to be achieved by employing a separate layer 2 address for each LAP in the system. To carry the LAP identity the address field is two octets long and contains a service access point identifier (SAPI), a terminal endpoint identifier (TEI) and a command/response bit. This address identifies the intended receiver of a command frame and the transmitter of a response frame. The address has only local significance and is known only to the two end points using the LAP. No use can be made of the address by the network for routeing purposes and no information about its value will be held outside the layer 2 entity.

The SAPI is used to identify the service that the signalling frame is intended for. Consider the case of I.420 telephones sharing a passive bus with packet terminals. The two terminal types will be accessing different services and possibly different networks. It is possible to identify the service being invoked by using a different SAPI for each service. This gives the network the option of handling the signalling associated with different services in separate modules. In a multi-network ISDN it allows layer 2 routeing to the appropriate network. The value of the SAPI is therefore fixed for a given service.

The TEI takes a range of values that are associated with terminals on the customer's line. In the simplest case each terminal will have a single unique TEI value. It is important that no two TEIs are the same and therefore the network has a special TEI management entity which allocates TEIs on request and ensures their correct use. The values that TEIs can take fall into the ranges:

- 0—63 non-automatic assignment TEIs which are selected by the user;
- 64—126 automatic assignment TEIs selected by the network on request;
- 127 a global TEI which is used to broadcast information to all terminals within a given SAPI.

The combination of TEI and SAPI identifies the LAP and provides a unique layer 2 address. A terminal will use its layer 2 address in all transmitted frames and only frames received carrying the correct address will be processed.

In practice a frame originating from telephony call control has a SAPI that identifies the frame as 'telephony' and all telephone equipment will examine this frame. Only the terminal whose TEI agrees with that carried by the frame will pass it to the layer 2 and layer 3 entities for processing.

13.5.3 Layer 3

The general structure of the layer 3 signalling messages is shown in Fig. 13.8. The first octet contains a protocol discriminator which gives the D-channel the capability of simultaneously supporting additional communications protocols in the future.

The call reference value in the third octet is used to identify the call with which a particular message is to be associated. Thus a call can be identified independently of the communications channel on which it is supported. This feature is particularly important in connection with incoming call-offering procedures on a passive bus arrangement since the channel is only allocated to the called terminal after answer.

The message type code in the fourth octet describes the intention of the message (e.g. a 'SETUP' message to request call establishment). A number of other information elements may be included following the message type code in the fourth octet. The exact content of a message is dependent on the message type; however, the coding rules are open-ended and in principle it is a simple matter to include additional information elements to satisfy any requirement which may be identified in the future. Currently the longest

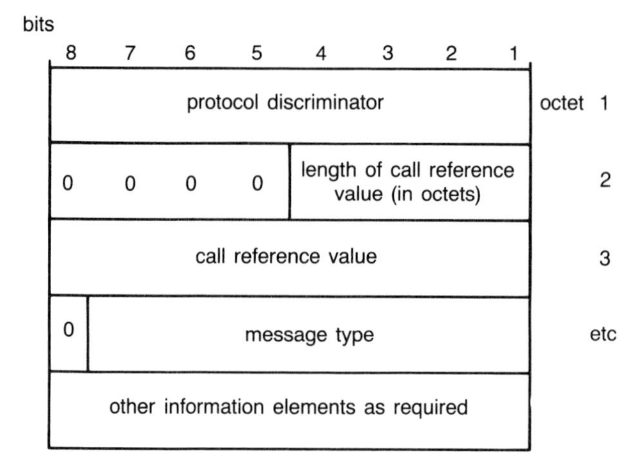

Fig. 13.8 Signalling message structure.

message type is the 'SETUP' message which contains 24 information elements, of which four are mandatory, and may be up to approximately two hundred octets long (if you include user-user information). Other messages may be restricted to just four information elements.

In order to make an outgoing call request, a user must send all of the necessary call information (i.e. called party number and supplementary service requests) to the network and this is done in the 'SETUP' message. Furthermore the user must specify in this message the particular bearer service required for the call (i.e. speech, 64 kbit/s unrestricted or 3.1 kHz audio) and any terminal compatibility information which must be checked at the destination. This information element may also be used to specify low-layer terminal characteristics such as data rate. Where applicable the non-voice application to be used on the call may be specified via the high-layer compatibility information element (i.e. Group 4 facsimile, teletex, videotex or slow-scan video).

BT currently uses the stimulus information element 'keypad' in the direction user-to-network to convey strings of IA5 characters for supplementary service control. This element conveys the results of key depressions at the man-machine interface presented by the terminal and through the use of the separators '*' and '#'; these character strings are arranged as facility requests which comprise a facility code representing the supplementary service followed by parameters to identify the various options selected. Thus a facility request may be represented in the following form:

<*><facility code><*><parameter 1>...<*><parameter *n*> <#>

The actual code arrangements for facility requests follow those defined by CEPT for multifrequency telephones and have been adopted in the CEPT recommendation for ISDN access.

The network response to facility requests will be generally in the form of a sequence of IA5 characters in a display information element or an in-band tone or announcement. The arrangement of IA5 characters within a display information element follows similar rules to the coding of facility requests. Use is made of the separators '*' and '#' so that sequences are machine-readable and suitable for display to a human operator.

The sequence of messages across the local network, between the calling terminal, the exchange termination and the called terminal, varies depending up on the features being invoked, but can be quite complex. The message sequence chart for establishing and clearing a basic call is shown in Fig. 13.9 and indicates when two similar terminals on an interface, with TEI = 5 and 8, respond to a 'SETUP' message.

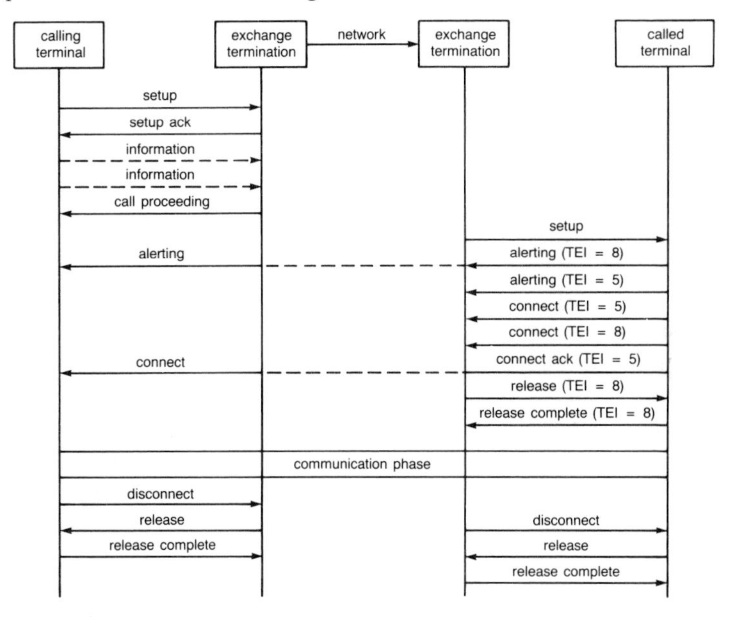

Fig. 13.9 Message sequences for basic call establishment and clearing.

13.6 BROADBAND ISDN

Attention is now being paid to how broadband services can be supported on a broadband ISDN. This is being particularly addressed by the European Community's RACE (research into advanced communications within Europe) programme, but it is also being considered by CCITT who in this current plenary period have agreed some 13 recommendations that have been put through the accelerated approval procedure.

The goal of the RACE programme is to make a major contribution to the objective of the introduction of integrated broadband communications (IBC) in Europe taking into account the evolving ISDN and national introduction strategies progressing to community-wide services by 1995. To achieve this the RACE programme is structured into three parts:

- Part I — IBC development and implementation strategies;
- Part II — IBC technologies;
- Part III — Prenormative functional integration.

Part I activities include the development of common functional specifications for IBC and include the specification of the signalling protocols to be used. This is being undertaken by a project which will be generating a detailed specification of the interface at the boundary between the public network and the terminal equipment or customer premise network. This interface is often referred to as the broadband user network interface (BUNI). The new protocols being defined for this interface will be carried transparently across the local network in the same way as for narrowband ISDN.

Fundamentally there is little difference between the signalling procedure requirements of a broadband service to that of the existing narrowband services. However, it is expected that, because of the bandwidth availability, multifunction terminals supporting several services simultaneously will be quite common and therefore enhancements in the repertoire of signalling messages defined for use at layer 3 will be required. Enhancements for controlling distributive services, such as TV and radio, will also have to be added.

The target network for supporting IBC, according to the CCITT Recommendations, will be based upon asynchronous transfer mode (ATM) techniques (see Chapter 12) and this may have considerable impact on the layer 2 protocols to be adopted by the signalling system. If the narrowband layer 2 procedures as defined in I.440/1 were adopted in the ATM environment, certain of its functions would be duplicated by the ATM adaptation layer, since this provides additional lower-layer multiplexing capabilities which may enhance or replace existing layer 2 multiplexing functions. It may therefore be desirable to modify the layer 2 protocol in order to suit ATM and take full advantage of it. For example, some functions which are carried out by frames in layer 2 could be carried out by the ATM cells. If layer 2 is optimized in this way, then it should be possible for layer 3 to operate independently of whether the network is ATM or STM. These issues have been discussed in the second document deliverable from the BUNI RACE project [7] and are currently being considered within CCITT.

13.7 CONCLUSIONS

This chapter has described the evolution of signalling systems in the BT local network. Digital access signalling systems have been defined for use in the narrowband ISDN trials and derivatives of these have been developed for use in private networks and to multiplexers located remotely from the host local exchange. A further DASS derivative has been defined which will also support telephony on remote multiplexers in an optical fibre local network. The DASS protocols are now established in the UK and their use has proliferated in the past owing to the lack of precise recommendations from CCITT. However, detailed CCITT Recommendations are now available for basic and multiline ISDN access and are being introduced into the UK. It is expected that the use of BT proprietary DASS will be phased out and gradually replaced by the CCITT Recommendations over a number of years.

Since the work on broadband ISDN protocols has started relatively early in Europe and in CCITT with the adoption of the first 13 recommendations on broadband ISDN in 1990, it is hoped that it will not be necessary for network operators to introduce proprietary systems. Further work on these and other recommendations is being carried out and it is hoped that detailed recommendations will be in place by 1992 in time for any widespread commercial demand for broadband services.

REFERENCES

1. British Telecommunications Network Requirement, Digital Access Signalling System, DASS No 1, BTNR 186 (1983).

2. British Telecommunications Network Requirement, Digital Access Signalling System, DASS No 2, BTNR 190 (1984).

3. British Telecommunications Network Requirement, Digital Private Network Signalling System, DPNSS BTNR 188 (1984).

4. Cattermole K W: 'Principles of pulse code modulation', Iliffe Books (1969).

5. Schickner M J: 'Digitalization of the junction and main networks', POEEJ, 74, pp 254-257 (October 1981).

6. ISO Data Processing — Open Systems Internconnection — Basic Reference Model. ISO TC97/SC16/537.

7. RACE Project 1044/UNI: 'Key Issues for the User Network Interface', RACE Deliverable Document (November 1988).

14

LOCAL ACCESS NETWORK MANAGEMENT

K J Maynard and P J Hawley

14.1 INTRODUCTION

The local access network can be considered as the first or final hop in the transmission network through which all network operator-supplied services pass. Most of these services pass through the local access network twice — once for each end connection. Only recorded information and enquiry services have a single pass through this network. For the purposes of this chapter, the local access network is regarded as starting at the transmission-terminating equipment in the serving local exchange and ending at the network-termination equipment at the customer's premises. Depending on the technology in question, the transmission-terminating equipment may be main distribution frame (MDF), digital distribution frame (DDF) or multiplexing equipment.

At present local access networks consist in the main of twisted copper pairs. However, other media are also in use, such as optical fibre, coaxial cable and radio. There is also a range of network topologies in use — star, tree and ring. If the network-terminating equipment in the customer's premises is also included as part of the local access network, then the overall technology used becomes an important aspect. For instance, twisted copper pair can be used to deliver simple analogue speech or digital signals up to (and in the near future beyond) 400 kbit/s. These technologies in turn are used to create and deliver the many services supplied to customers.

In order to increase the bandwidth and flexibility for customer service delivery, ever more complex technology is being developed. The large investment needed to develop this is inevitably concentrating the source of such technology in a small number of very large multinational companies. This means that all network operators in future could potentially have the same range of technology with which to deliver customer services. It is therefore the efficiency with which network operators manage the technology resources available to them that will distinguish them, while the technology itself will be a secondary issue in gaining profitability. Market share will be determined by an operator giving customer satisfaction and value for money.

Technology variants have been evolving, and are continuing to evolve, providing network planners a diversity of solutions for any one customer application. These local access network technology variants can be regarded as an expanding portfolio of piece parts. It is these piece parts that a service provider and network operator must use and manage in order to meet customer requirements efficiently, flexibly and economically.

14.2 TECHNOLOGY VARIANTS

When discussing the management of the local access network, it is useful to establish what some of the technologies in use are, and what management opportunities they provide. This chapter does not attempt to be a definitive reference of all technologies in use or proposed for the local access network. It should, rather, provide an insight into some of the technological possibilities and their individual management considerations. The technologies have been chosen in order to highlight the main management issues associated with topology, media and analogue/digital systems.

A number of management considerations are common to all or many of the technologies. These include:

- the configuration, capacity and geographical location of underground plant — manholes, joint boxes and duct;
- the configuration, capacity, physical make-up, transmission characteristics, allocation and geographical location of cables;
- potential serving domain of DPs;
- test access;
- inventory and performance aspects of element types.

14.2.1 Single pair of metallic wires

This is the most basic of all technologies and the backbone of many more complex technologies. A single pair of wires is dedicated to an individual customer circuit (Fig. 14.1). It has a star topology, with the possibility of multiple flexibility or cross-connection points between the MDF and the NTE at the customer's premises. The NTE for this technology is a purely passive device. This technology can be used to deliver POTS, telex or private-circuit services.

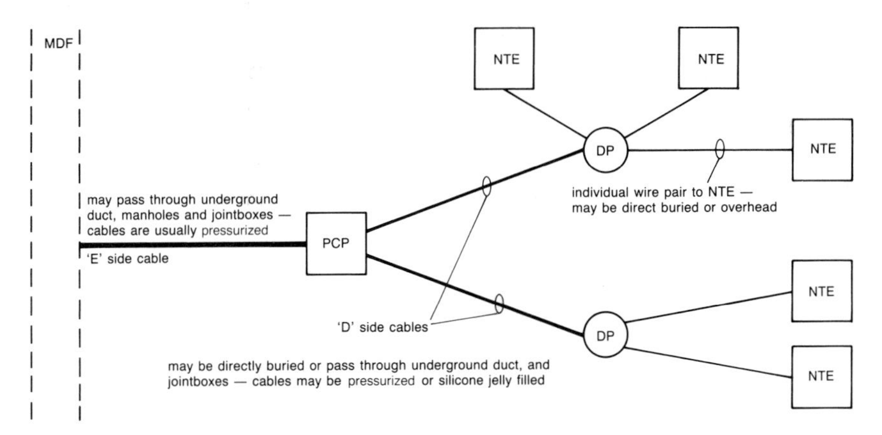

Fig. 14.1 Metallic pairs.

Specific management considerations include:

- the configuration, capacity, pair allocation and geographical location of cross-connection points — MDF, PCPs and DPs;
- determination of transmission loss parameters given line length;
- monitoring of cable pressure alarms.

14.2.2 Metallic pairs and modems

This technology uses the metallic pairs to transport data in analogue form (Fig. 14.2). The modem is used as the conversion unit from analogue to digital transmission. If the technology uses a single metallic pair then it may be connected to a standard analogue line card in the local exchange to provide PSTN access. It may also be used for low data-rate private circuits. The use

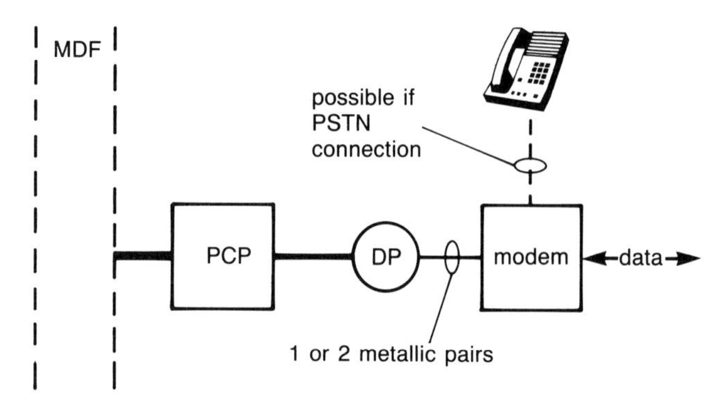

Fig. 14.2 Metallic pairs and modems.

of two metallic pairs is another option for providing higher data-rate private circuits.

Specific management considerations include:

- all those involved with single-pair metallic-wire technology;
- remote soft configuration of modem unit;
- test access and diagnostics;
- event handling.

14.2.3 Pair gain DACS 1

The digital access carrier system no 1 (DACS 1) is a two-channel technology providing analogue interfaces to the exchange and the customer, but using digital transmission between the units (Fig. 14.3). The medium used is a standard single metallic pair. The NTE is a purely passive device. This technology can be used for access to the PSTN or for analogue private circuits.

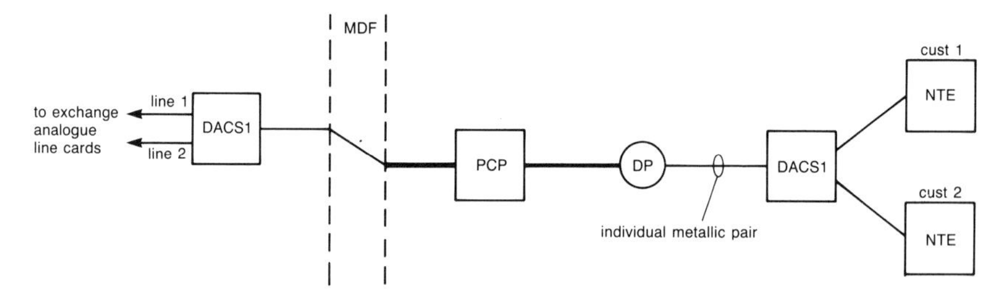

Fig. 14.3 Pair-gain DACS 1.

Specific management considerations include:

- all those involved with single-pair metallic-wire technology;
- configuration of multiplexing units to single metallic pair;
- logical/physical addressing of multiplexing units;
- planning/feasibility given transmission loss parameters;
- basic line test facilities of metallic drop to customers, with the provision of mimic impedances for exchange line testers, to enable the correct repair staff to be dispatched;
- monitoring of digital transmission path.

14.2.4 Fibre access network (FAN)

With this technology, the customer's premises are connected to the local exchange by optical fibre. This provides an access mechanism for a number of services (Fig. 14.4). The connection comprises a number of fibres to allow diverse routeing for reliability.

The equipment at the customer's premises comprises optoelectronics, high order multiplexers, line transmission protection switching — common equipment which together comprise the network service module (NSM).

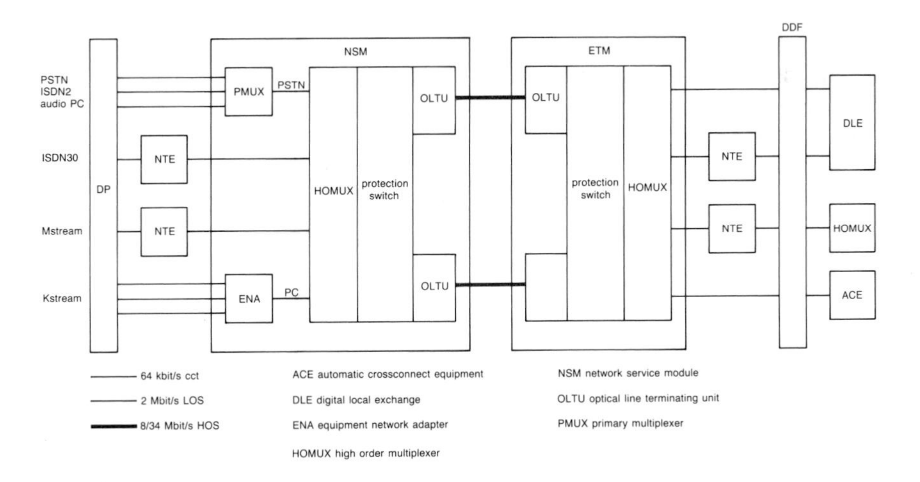

Fig. 14.4 FAN configuration.

The primary multiplexers consist of a series of line cards which provide the interface to the customer (via the building distribution frame) for the services required. These 64 kbit/s inputs are then multiplexed together for forward transmission via the high-order systems to the exchange serving site.

At the serving site the HOSs are de-multiplexed back down to 2 Mbit/s LOSs. These are forwarded, via the bearer transmission network, to the service networks (e.g. PSTN, X-stream or ISDN).

Specific management considerations include:

- configuration and allocation of elements — MUX, line card, optical end, NTE type;
- state control of elements;
- automatic event reporting.

14.2.5 Optical fibre, switched broadband

This is the technology employed in the BT switched-star cable TV network (Fig. 14.5). The only extant installation is located in the City of Westminster [1,2]. Video channels are multiplexed together and transmitted down optical fibres, four video channels per fibre. Some video channels are broadcast via the hub-site, which performs fan-out to the wideband switch points (WSPs). Other video channels are sent via dedicated optical fibres to the WSP. The WSP converts the optical signals to electrical signals and feeds them to the inputs of the video switches. When a channel is selected by the customer, the appropriate switching is performed by the WSP to provide the required signal. The final link to the customer is via small-bore coaxial cable. Text generators in the WSP allow for customer interaction with information databases held at the headend.

Specific management considerations include:

- configuration of subsystem control and data links;
- configuration and allocation of subsystem elements;
- serving area of WSPs and DPs;
- monitoring and control of subsystem and element states;
- automatic event reporting for subsystem elements including signal strengths;
- performance aspects of element types;
- remote update of WSP database;

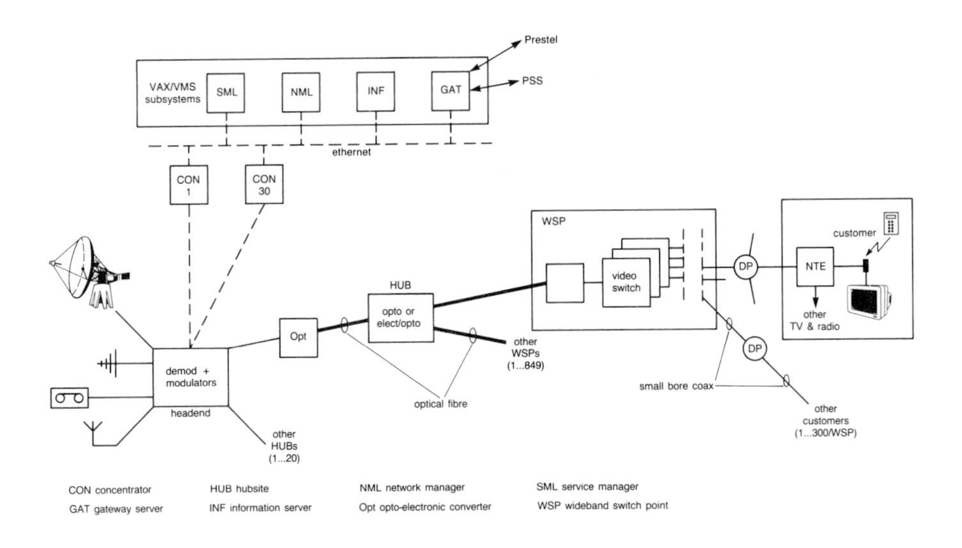

Fig. 14.5 Switched-star schematic.

- passing information to control customer access to video channels and/or data services.

14.2.6 TPON (house/business)

TPON (see Chapters 9 and 11) is a digital technology which uses optical fibres in a 'tree' topology and employs TDM/TDMA techniques (Fig. 14.6). The optoelectronics are used to separate the multiplexers from their line cards. This allows the MUX line card to be remotely located in the customer's establishment. For residential or 'house' TPON the NTE contains only a single line card. However, for business applications, the NTE can contain a number of line cards, each of a different type. The services provided by this technology are dependent upon the type of line card used, and the bit capacity allocated.

Specific management considerations include:

- potential serving domain of secondary splitter;
- configuration and allocation of elements — MUX, line card, bit capacity and position within TDMA, optical end, NTE type;
- state control of elements;
- automatic event reporting;

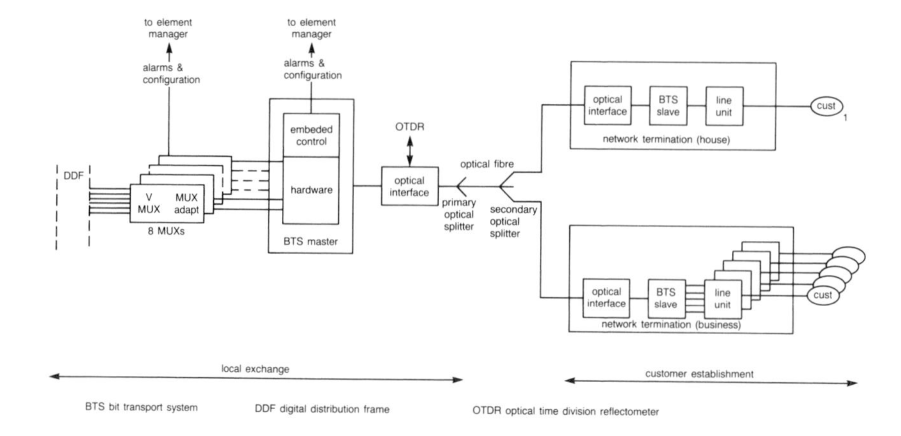

Fig. 14.6 TPON schematic.

- attribute set-up of bit transport system (BTS) slave;
- test access, including optical time domain reflectometry.

14.2.7 Radio, point-to-point

As Fig. 14.7 shows, a multiplexer is remotely located at the customer establishment and linked by the use of a line-of-sight radio system. It is currently only used to provide megastream.

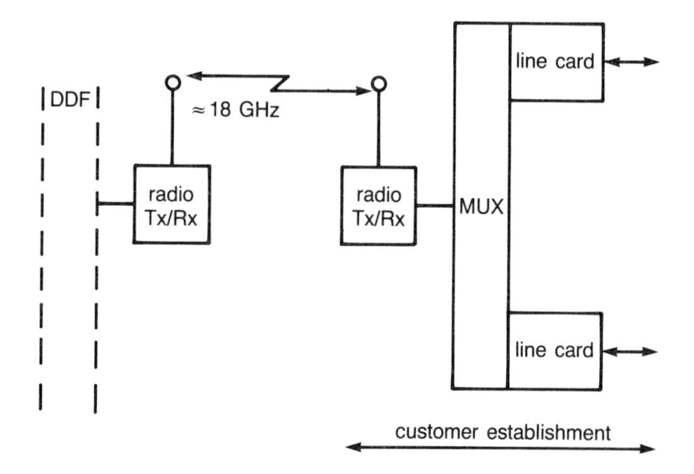

Fig. 14.7 Radio schematic.

Specific management considerations include:

- geographical location of radio units and multiplexer;
- planning/feasibility by — determination of line of sight, avoidance of interference to or from other systems, signal strength;
- allocation of frequency;
- configuration of multiplexer;
- line card types, locations and allocation.

14.2.8 Mobile GSM

GSM (Group Speciale Mobile) uses digital cellular radio techniques to provide connection from portable or mobile telephones into the PSTN. The specification is common across Europe and will allow suitably authorized mobile telephones to access the system at any point in its coverage area across Europe. The system will allow data calls to be made as well as voice calls and also provides a paging service.

Specific management considerations include:

- base-station site planning and cell coverage area determination;
- cell frequency allocations;
- GSM network planning and PSTN access point determination;
- mobile authorization — both on first subscription and per call;
- call accounting, charge determination and assignment;
- system performance monitoring and refined coverage area determination;
- system growth planning and provision.

14.2.9 Cordless DECT

The DECT (Digital European Cordless Telephone) standard provides for low-power digital portable radio telephones providing connection to a base station over a short range. The base station may be located in the home, connected to a standard telephone line, or it may be located in a public place, e.g. a railway station or motorway service area. In this case only outgoing calls from the portable can be made. The technology can also be used to provide a PBX service.

Specific management considerations include:

- base-station site planning and coverage domain determination;
- DECT public access network planning and PSTN access point determination;
- portable authorization on first subscription;
- call accounting, charge determination and assignment;
- system performance monitoring;
- system growth planning and provision.

14.3 MANAGEMENT CONSIDERATIONS

The local access network is unique in at least one respect of its management requirements — the need to determine the potential network access points or points of sale with respect to a geographic identifier for the customer's address, the postal address or map reference co-ordinates.

It is likely in the near future that a given customer location can be served by more than one technology. Management systems must be able to:

- identify the technologies that are capable of providing the service required by the customer;
- select between the different technologies on the basis of cost, reliability, performance and availability.

In the event of insufficient capacity, the management system should also be capable of managing the provision of customer service via a relatively more expensive technology in order to satisfy an urgent requirement. This would be followed by supervising the planning and transition (at a later date) to a more cost-effective technology. For example, it may be possible to provide a customer's circuit quickly via a radio-based technology, whilst planning a change to a fibre-based technology. It may also be possible to upgrade a metallic-pair-based technology as a result of opportunities presented by the introduction of pair-gain systems.

The impact on the planning process of the presence of multiple alternative networks is likely to be considerable.

In addition to the management considerations of the various technologies, future network and service management systems should also consider the issues of facilities required by the network operator.

- Customers should be provided with a single point of contact for all their enquiries and calls for assistance. It should be possible once the customer has requested assistance to be able to respond rapidly to the customer, regardless of topic. Comprehensive workforce-management facilities should be available to support field installation and maintenance activities.
- Centralized network management centres should have the ability to test circuits remotely. By employing the techniques of routining, alarm gathering and 'hot-spot' analysis, these centres should also be capable of detecting any deterioration of service, ideally before, but at least as soon as, a customer experiences difficulty. It should also be possible for faulty circuits to be taken out of service automatically and the network management centre informed. Auto-transfer of service to back-up circuits should be undertaken where the technology in use allows for diverse routeing. Wherever possible maintenance tasks should not degrade the quality of service, and field maintenance activities should be minimized.
- Strategic objectives for management systems should ensure that the system design is evolutionary and expandable for the easy addition of new operational facilities and the management of new types of technology. In order to achieve this the system design should conform to an overall telecommunications management architecture employing international and open standards as these emerge. It will be this system design which provides a competitive edge for the network operator over its rivals.

The ability of management systems to meet these objectives will depend on the underlying network technology in use, in particular the degree of built-in intelligence for monitoring and control, and equipment redundancy for fault recovery.

Network operators and the technology providers of new equipment or services often have different perspectives of management requirements. A technology provider would normally wish to sell equipment and an associated management system to cover all aspects of its use, from service management through to the operation of the network elements. The technology provider perceives a complete turnkey system as the best way to market his product, so he produces a 'vertical slice' (Fig. 14.8). This will contain an integrated management capability, but for this particular product only. The network operator, on the other hand, will have structured (or planned to structure) his operation in a layered fashion, in 'horizontal slices' to provide a single customer-facing point of contact for the 'one-stop shopping' which is considered to be the key to fostering good relationships with customers.

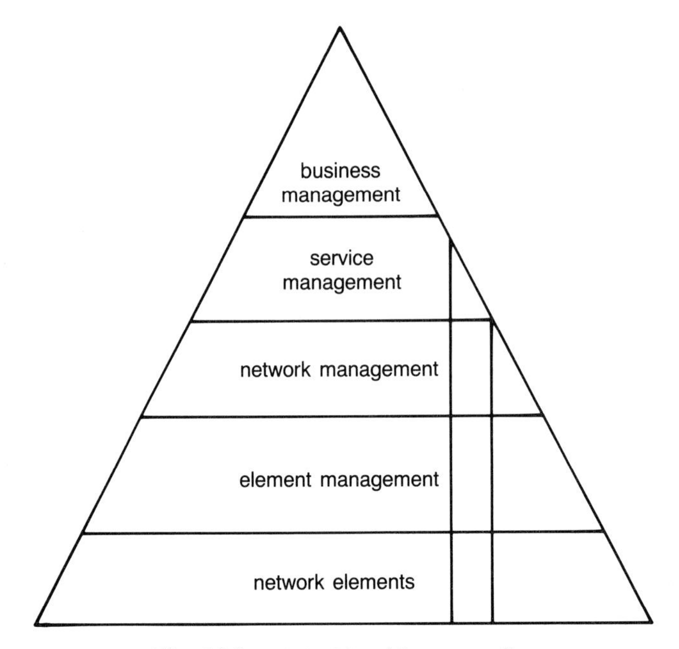

Fig. 14.8 CNA-M architecture outline.

The different perspectives of network operators and technology providers could both be addressed in the same system package from a technology provider. This would provide interfaces at the business level, service level and network control level to facilitate integrated working to satisfy the network operator, whilst still providing a complete turnkey operation for those who require this.

The different technologies described above provide the network operator with many varied opportunities and challenges. With the different technologies available, and an integrated management solution with customer access, it is likely that in the future the customers themselves will be able to configure new services without the intervention of the network operator.

To reduce the provision time of customer circuits, it will be necessary to ensure that as much of the time-consuming preliminary work as possible is carried out in anticipation of a customer order. With the growth of competition to supply even basic telephony service, improved efficiency must be given more attention. In the USA, operators are increasingly employing the 'warm dial-tone' concept, where new buildings are pre-wired and telephone sockets connected through to the exchange. The new resident plugs in a phone purchased from the local store and is connected directly to the network operator's customer service office. The new sale can be initiated immediately with the required record changes, billing function, directory number allocation and service provision taking place automatically. To achieve this, a fully integrated set of management functions must be in place.

14.4 MANAGEMENT SYSTEMS ARCHITECTURE

It is important that network-management systems develop to overall architectural standards if such systems are to inter-work and to provide integrated network management and service administration across the network at all levels. Recognizing that the local distribution network is an integral part of the overall telecommunications network, it is important that management of the local network evolves within such an overall architectural framework. The following outlines a general architectural model considered from a local network viewpoint.

The architecture for telecommunications network management is defined in terms of a hierarchy of five layers [3], as shown in Fig. 14.8. Starting from the bottom layer, the network is partitioned into network elements which are treated as distinct network entities from a management viewpoint but which co-operatively provide a service to the customer.

As shown earlier, the local access network is made up of a number of managed element types, e.g. network termination units, MUXs, cables and DPs. Each network element has an associated element manager. This may, in practice, be a separate processor linked to the element, or software embedded within the element. An element manager would normally manage multiple instances of an element.

The network-management layer, in turn, supervises the various element managers. This is the first layer where the management relationships amongst elements are co-ordinated to provide overall network and technology supervision. The interface between an element manager and element may be specific to that element-type. The element manager 'hides' equipment variability from the higher network-management layer. For example, there may be different types of modem in the network, all providing the same basic service, but requiring different control streams to configure them. The differences would be confined to the element manager level.

This 'layering', applied to the components which form the local access part of the network, is exactly the same as that applied to the switching and transmission parts; the architecture is common.

Workforce-management facilities would be linked to this level to ensure that maintenance and repair activities are co-ordinated with network management.

Thus the higher-level interface between element managers and network management is independent of particular equipment implementations. This application of the design principle of 'layering' achieves open network management whereby the network control level can develop independently of changes in underlying equipment.

The interface between network and service management levels is defined such that service management need not know about the physical details of the network and need be concerned only with administration of the services that the network supports. Service management should have sufficient interaction with the network to allow comprehensive development of 'single point of enquiry' and 'one-stop shopping' administration facilities. Refining this general model from a local network viewpoint, the elements making up the overall telecommunications network fall into two groups — those within the 'core' segment and those within the 'access' or 'local' segment of the network, with their respective element managers. Co-ordinated supervision is necessary across the access/core segment boundary, and would be implemented at the network-management level, as represented by Fig. 14.9.

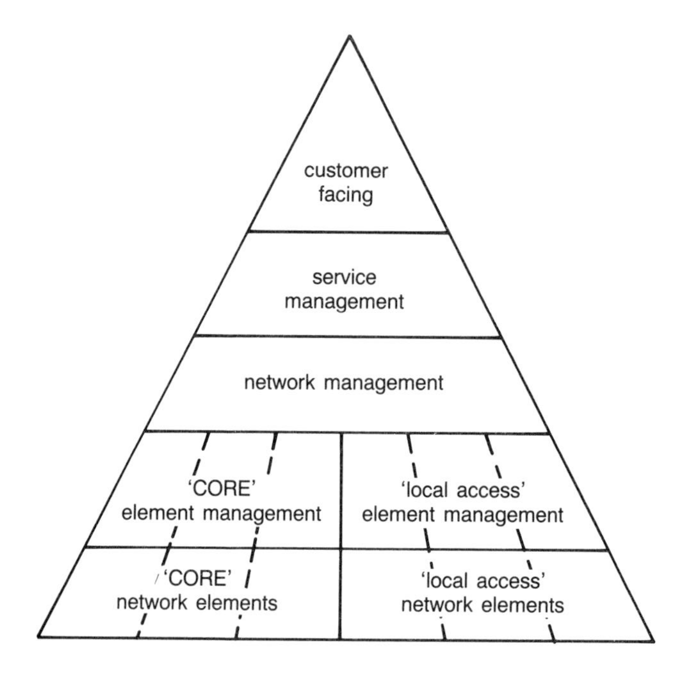

Fig. 14.9 CNA-M architecture — local access perspective.

Traditionally, different services have been carried on physically separate bearer circuits, particularly in the local network. The management of the services and of the networks which bear them have often been common. In future multiservice networks will be normal, and services and networks will have to be managed separately if the desirable layering structure is implemented across the network operator's business, but with agreements made between the service and network management units to ensure that the overall quality of service requirements of the customer are fully met.

A model for technology-independent management of the local access network based on CNA-M is currently under consideration. The model attempts to prevent diversification, or duplication, of effort and the urge to re-introduce vertically segmented solutions when new technologies and services are developed.

14.5 FUTURE WORK

The model achieves technology independence through the functionality of the layers, the data contents of the layers, and the data definitions and relationships. The model defines the local access network using object-oriented techniques and the use of the OSI/network management forum object library. Data is passed between the functional categories and the layers using the OSI common management interface protocol.

The model uses the standard CNA-M network control layer functional categories — configuration, event handling, performance, access and security, planning, finance and resource management — to achieve conformance with the architecture and to encourage reusability.

It should be noted that changes will have to be made to the current management systems and processes if substantial benefits towards managing the local access network in a technology-independent manner are to be made.

It may well be that, in future, the business case for the introduction of new proprietary technologies will include a requirement for the product to be capable of control using the established management model, to ensure that BT maintains a coherent product set.

REFERENCES

1. Boyd R T: 'Service administration and network management on the BT switched-star network', BT Technol J, 5, No 3, pp 32-43 (July 1987).

2. Boyd R T: 'Interactive service development on the BT switched-star network', BT Technol J, 5, No 1, pp 57-53 (January 1987).

3. CNA — RD0035: Part 1:1,00: 'Management Interface Architecture: Introduction and Model'.

Part Five

Future

15

FUTURE PROSPECTS FOR THE LOCAL NETWORK

J Stern

15.1 INTRODUCTION

In this concluding chapter the major trends discussed earlier in the book are summarized and the future shape of the local network likely to emerge over the coming 10-20 years is discussed.

The vision of an all-fibre telecommunications network delivering an ever-increasing variety of services to both business and residential customers has been the subject of much speculation and prediction for more than a decade. However, although optical fibres are now used routinely for longer-haul transmission, extending fibre to provide the final link to the customer is proving to be a tougher challenge than originally expected. This chapter summarizes the thinking behind the more sophisticated and commercially based approaches now under investigation worldwide to achieve large-scale deployment for business and, eventually, residential customers.

By the early years of the next century the local network will be based on a mix of fibre, radio and copper technologies and new technological developments for the latter two media are also considered. The relative proportion of each transmission medium in the network, and in particular the extent of fibre penetration, will depend not only on technical capabilities and economic performance but on regulation, competition and the types of service that customers will actually want. The interplay between these complex factors is discussed and likely scenarios described.

15.2 CURRENT NETWORK SITUATION

Digitization of the BT trunk (inter-city) network was completed in March 1990 with some 398 000 fibre km representing the dominant transmission medium. Digitization of local telephone exchanges is now proceeding apace with some 4500 digital exchanges expected to be in operation, serving 60% of customers, by 1992. Optical fibre is again the dominant transmission medium in the junction network linking these exchanges with 493 000 fibre km installed by March 1990. By contrast the local network, as in all other advanced countries, remains overwhelmingly dominated by traditional copper-pair technology. By September 1989 some 36 million copper pairs were installed in the UK network (23.3 million actually connected to customers). Today, however, although over one million kilometres of fibre have been installed in BT's network, only 10% has penetrated the local loop.

The key barrier to widescale deployment of fibre in the local loop is cost. Early ideas for the introduction of fibre were based on the simplistic assumption that the widescale introduction of broadband services — such as cable TV, pay TV, dial-up video services, and high-definition television — would justify the extra cost of the optical fibre over the copper pair. In reality, however, regulatory restrictions and the uncertain demand for new entertainment TV have led to a much more sober assessment of the likely timescales for the introduction of these services.

In the meantime telecommunications companies around the world are continuing to service new growth in their network by installing more copper pairs, technology which could rapidly become obsolete when the broadband revolution eventually arrives. Much of this growth is for the business sector where the demand is for higher-quality, more responsive networks for the delivery of relatively conventional telephony and data services. Customers in particular are requesting the delivery of tailored packages of services (telephony, private circuits, switched data, etc) that can be dynamically varied to keep pace with their changing needs — preferably without the need for formal intervention by the telephone company.

These developments have led to a review in the priorities for technological development in the local loop leading in particular to a new breed of optical system design aimed at delivering existing narrowband services to the customer in more flexible ways — but capable of being subsequently upgraded to supply broadband services when the market demand actually arises (section 15.3). The continuing importance of the copper loop is also more fully recognized than before and a resurgence of technical development is under way in this area (section 15.4). Mobile telephony has emerged as a major new focus for customer interest and this also will be a key factor in shaping the future local network (section 15.5).

15.3 FIBRE SYSTEMS

15.3.1 Fibre architectures

Various fibre architecture options have been considered for local networks including star, bus and ring architectures, of which stars have been most widely considered since they map directly on to the existing duct infrastructure used for copper pairs.

Stars can be single stars, double stars or, possibly, star buses (see Fig. 15.1). The first of these is a direct analogy of the existing copper network where a fibre runs directly from the exchange to the customer. This architecture is relatively expensive because of the need for a dedicated fibre and two optical sources and receivers for each customer. Advantages, however, are a high level of privacy for each customer, simple fault location

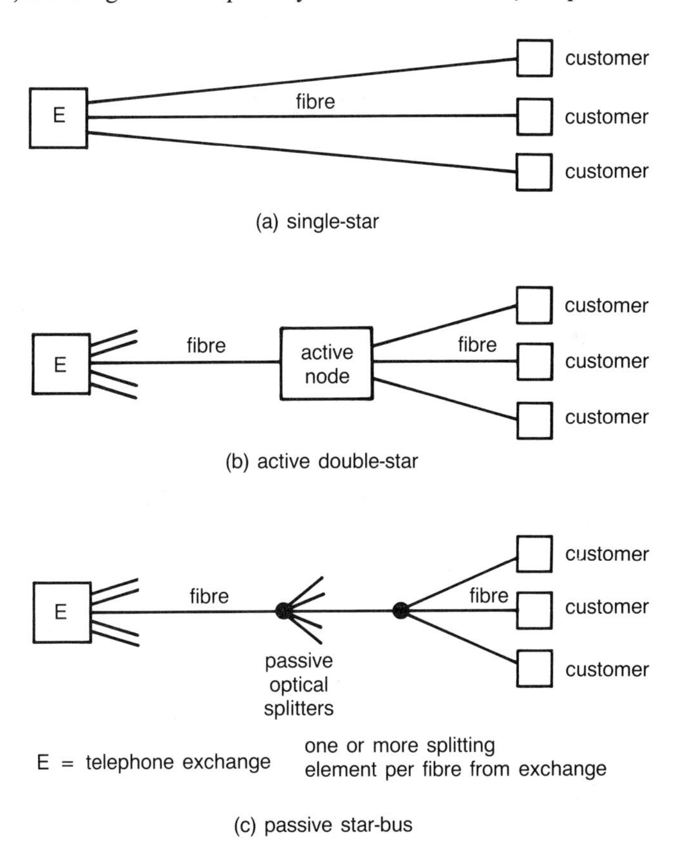

Fig. 15.1 Fibre architectures.

and future-proof service evolution. Single stars are best for large business customers requiring large capacities (e.g. 100 telephone lines or more) where the relatively high provisioning costs can be justified.

Double-star topologies can be either active, containing intermediate electronics, or passive, containing optical splitters (Fig. 15.1). Essentially they allow the cost of the network between the exchange and the active or passive splitting node to be shared amongst a number of customers. These approaches are the most promising for allowing fibre to be economically extended down to small- and medium-sized business and residential customers. The main advantage of the active approach is that the final link to the customer is dedicated and can make use of low-bandwidth, low-performance optical devices. On the other hand there are problems with locating, powering and operating remote electronics in the field and with the high up-front costs involved. The key advantage of the passive approach is that initial network costs are particularly low with the majority of the cost being deferred until revenue-earning customers are actually connected. On the down side, the customer requires higher-bandwidth, better-performance optical devices and some method must be implemented to ensure privacy and integrity of information, since the signal is broadcast to all locations.

15.3.2 UK system developments

Single-star systems have been developed and deployed in most major advanced countries for the provision of telephony and other narrowband services to large business customers. Most of the fibre installed in the UK local network is for this type of application and has been installed in the financial business districts of the City of London and London docklands [1]. The essential features of this type of system are dedicated fibre links operating, typically, at 8 Mbit/s and flexible, intelligent customer multiplexers able to cope with the considerable growth and churn of large business traffic.

Double-star systems aimed at smaller business and residential applications are currently at the field trial stage. Trials are being mounted around the world to gain experience in the practical issues involved in future wide-scale deployment and to compare and contrast the technical options available. BT has developed two experimental systems, one active double star and one passive, and is trialling both at Bishop's Stortford in Hertfordshire [2]. This venue has been chosen to offer a representative combination of new and old housing developments, greenfield sites and business parks with mixtures of overground and underground cabling. The range of services being offered in the trial are combinations of telephony, 18-channel cable television, 16 stereo audio channels and videotext services.

The active double star system, known as BIDS (broadband integrated distributed star), shown in Fig. 15.2, has been more fully described in Chapter 8. Telephony traffic is carried at a bit rate of 140 Mbit/s to a remote access point within the network serving some 240 customers. An electronic demultiplexer then selects the channels for each customer which are then remodulated on to an individual fibre that runs to the customer's premises. Broadband programme material (18 TV channels, 16 stereo radio channels and videotext) is also transmitted to the remote access point via a separate fibre using electronic frequency modulation techniques. An electronic switch at the remote node selects the material chosen by the customer which is then also modulated on to the same individual fibre link as the telephony.

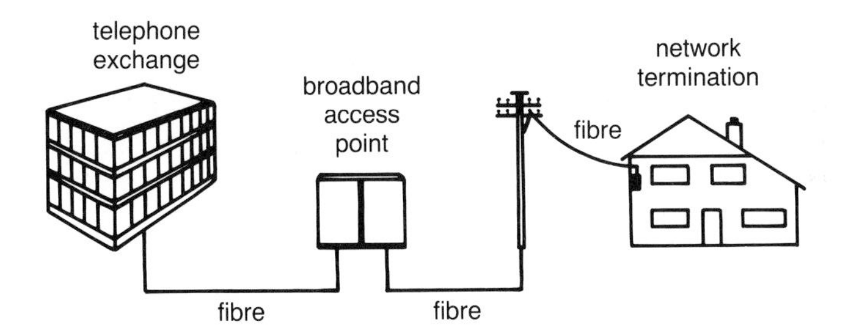

Fig. 15.2 The active double-star system (BIDS).

The passive system, known as TPON (telephony on a passive optical network), is shown in Fig. 15.3 and is also more fully described in Chapters 4, 9 and 11. A single fibre emerging from the local exchange is fanned out via passive optical splitters at suitable points in the network to feed a number of customers (typically up to 32). A time division multiplexed (TDM) signal at 20.48 Mbit/s is broadcast from the exchange to all customer terminals. Each customer's terminal is allowed by the control protocol to access only the channels intended for that destination. In the return direction, data from the customer's terminal is inserted at a predetermined time to arrive at the exchange within an assigned time slot. The heart of TPON is the time management of the time division multiple access (TDMA) system and is implemented by a ranging protocol that periodically determines the path delay between the customer's terminal and the exchange and updates a digital programmable delay in the customer's equipment. The system has a capacity of 240 telephone channels (each at a bit rate of 64 kbit/s) and these can be flexibly allocated to any of the customer's terminals in line with changing customer demand. It is also possible to organize different types of traffic (e.g. public switched circuits, private circuits, kilostream services, etc) to return to the exchange in such a manner that switch ports can be efficiently loaded.

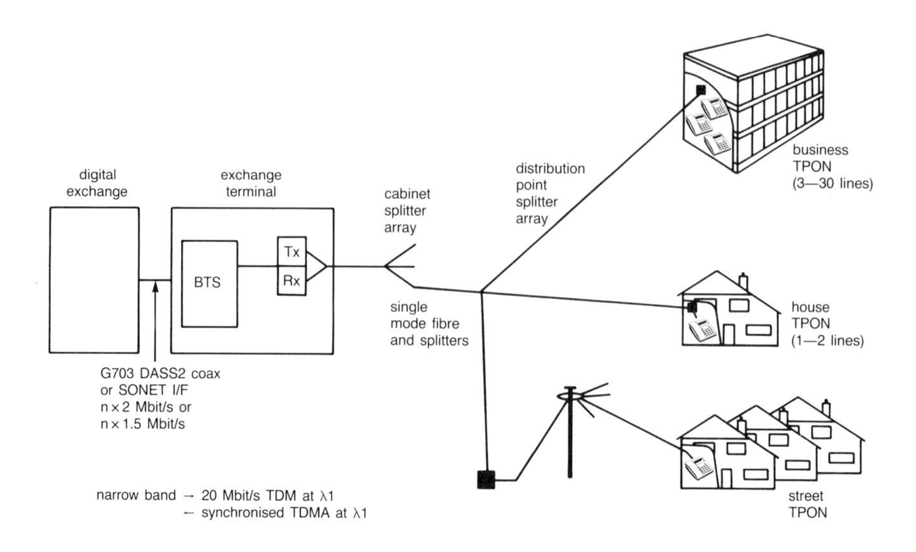

Fig. 15.3 The passive double-star system (TPON).

The TDMA approach involves automatic monitoring of the time of flight and amplitude of the optical pulses to accuracies of 5 ns and 0.5 dB respectively. This allows powerful diagnostic and maintenance features to be built into the system so that, in most instances, faults can be pinpointed without the need for time-consuming field investigation. Some impending faults (e.g. gradual customer laser failure) can be detected and scheduled for repair before the customer is aware of the problem.

Three forms of the TPON system (Fig. 15.3) are being trialled at Bishop's Stortford — house TPON providing fibre-to-the-home (FTTH), business TPON providing fibre-to-the-office (FTTO) and street TPON providing fibre-to-the-curb (FTTC) with copper pairs for the last few metres to the home. The economics of both business TPON and street TPON are significantly more favourable because the cost of providing the fibre sytem is shared over a number of telephone lines.

The TPON system operates at the optical wavelength of 1300 nm. By adding other wavelengths it is possible to introduce new services in the future without needing to add more cables. These could be broadband services such as cable television or they could be business services such as high bit-rate data, video conferencing or video telephony. Several different modulation schemes have been demonstrated to carry broadband services including AM, FM/FDM and digital (see Chapter 10). At Bishop's Stortford FM/FDM techniques are being used to carry 16 TV channels on a single additional wavelength to some 30 BT customers. This approach allows the same set-top box as used in satellite TV reception to be used to demodulate the signals in the home.

As experience from Bishop's Stortford and other studies accumulates, BT is favouring the passive approach as the most promising way forward for achieving wide-scale deployment of fibre systems. This is because economic deployment of the BIDS-type system is very dependent on a well-defined and high-penetration demand for broadband entertainment TV services to the home. The PON methodology is much more suited to a low-cost telephony entry approach with flexibility for providing (unknown) broadband upgrade over a period of time.

15.3.3 World scene and standards

Elsewhere in the world, notably the USA, active double-star systems are being trialled in FTTC form (see Fig. 15.4) [3] purely in the first instance for the delivery of telephony and other narrowband services. These now represent the main competition to the passive approach. However the low up-front costs and flexible service delivery characteristics of PONs make them particularly well-suited to the narrowband services entry situation — especially in the European context where there is little experience in deploying large remote active nodes within the local network. Following BT's pioneering role, a number of PON trials are under way within Europe. Germany (see Chapter 2), which now faces the need to renovate completely the very poorly equipped network in the east of the country, has seven concept trials of various types coming on stream. France has already trialled a PON system [4] and, in Holland, the Dutch PTT are deploying a trial system developed from BT's TPON [5]. Spain is also mounting a trial. In North America the PON approach is also receiving enthusiastic support from some of the major telephone companies (see Chapter 3) and trial systems are now being deployed by six out of the seven regional Bell operating companies. In Japan, Nippon Telephone and Telegraph have recently announced field trials planned for

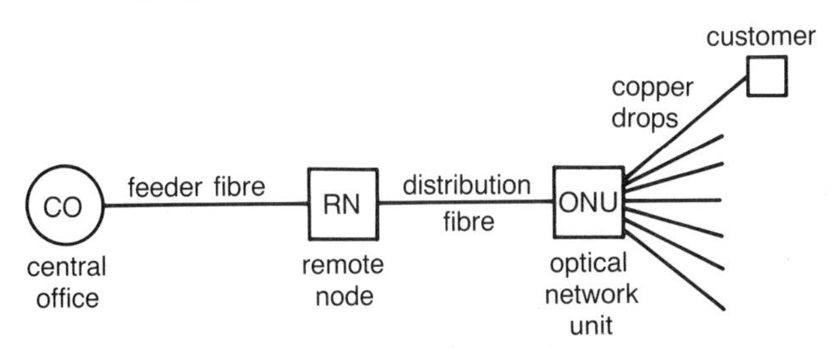

Fig. 15.4 FTTC active double-star system.

1992 (see Chapter 1) which will include PON systems targeted at FTTH deployment. Interestingly these NTT systems will attempt to achieve economic viability in the most testing of situations by innovating new narrowband services to deliver to the domestic customer.

Although BT has, experimentally, adopted an FTTH approach for delivering broadband entertainment TV to customers in Bishop's Stortford, it is likely that FTTC architectures will be used in the nearer term by network operators that have viable early broadband markets. FM or AM modulation format will be preferred to digital until market size becomes large enough to warrant development of the appropriate codec standards and VLSI electronics — or until TV sets with digital inputs become the norm. AM has the major advantage over FM in that no cumbersome customer set-top box is required — allowing normal use of the home remote control. Until recently, however, the drawback with AM has been the very restricted optical power budget available for implementing optical splitting. This has meant that initial system designs, such as Raynet's [6], have either featured very low split for the broadband PON (e.g. four ways) or, alternatively, no split at all with the broadband being carried over a separate fibre installed at the same time as the telephony entry network. Recent developments in optical amplifiers, however, both at 1550 nm (erbium-doped fibre devices) [7] and at 1300 nm (Pr^{3+}-doped fluorozirconate fibre devices) [8] look set to allow PONs to be engineered with fully adequate split levels for the first time. The very recent breakthrough in 1300 nm devices could be particularly significant in the future since linear semiconductor laser devices are available at this wavelength at reasonable cost.

15.3.4 Standards developments

Standards discussions within the international bodies are now becoming increasingly important in the quest for cost-effective fibre entry systems. The market of one telephone company — especially if the initial target is, say, FTTC for new growth — is necessarily limited and significant cost/volume advantages are potentially possible by making common cause with others. The difficulty however is to determine what can be the initial standardization aims, given the relatively immature state of development of FITL systems and the diverse application areas (FTTH, FTTO, FTTA (fibre-to-the-apartment), FTTC) and service scenarios envisaged by various operating companies around the world.

In general there are four locations in the network where future standardization may be appropriate (see Fig. 15.5). These are the switch interface to the exchange termination (ET), the interface to the customer's equipment and the optical interfaces to the passive distribution network (which could be a point-to-point link or a PON). Discussions are under way in Europe (ETSI) on the first of these (important to allow independent procurement of switch and local access network) and, for the second, customer equipment standards are already covered by various existing and pending national/international standards (largely stemming from the existing copper network).

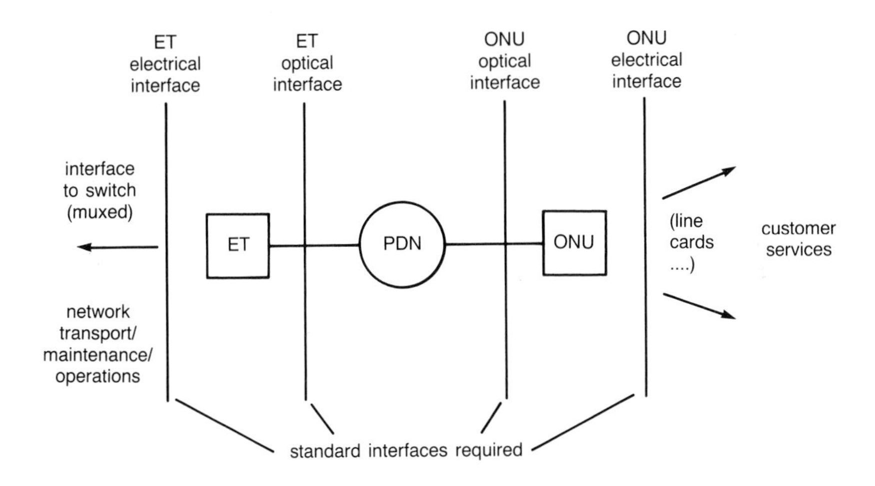

Fig. 15.5 The general structure of the fibre access network.

In the longer term optical interface standards of the SDH type will be desirable to permit the separate procurement of ET and ONU equipment. However this is premature at present when the pros and cons of competing system offerings are not yet fully understood. An interim approach, as developed by Bellcore, has been to establish a guiding framework within which manufacturers can develop competing ET and ONU equipment designs to operate over the PDNs of various network operators [9]. Such guidelines avoid the detailed specification of system bit rate and protocol but address such key parameters as maximum split size, system range, ONU modularity/size/capacity, operating wavelength, generic OA&M requirements and broadband evolution requirements.

The overall cost of fibre systems, particularly PONs, is heavily influenced by the cost of the ONU and, within this unit, optoelectronic components form a significant proportion of the total. An early priority therefore should be to standardize specifications for these devices to encourage the availability of low-cost components for use by all systems manufacturers.

Beyond these initial ideas for standards a further possibility to achieve lower systems costs may be to consider the subdivision of the PON system into the core portion, concerned with basic 'bit transport', and a service shell responsible for interfacing particular services to the core (Fig. 15.6). Thus the concept would be of a standardized interface (or backplane) between the core and shell. The core could then be deployed against diverse application areas and service requirements with specific service shells being locally developed to meet the needs of individual network operators.

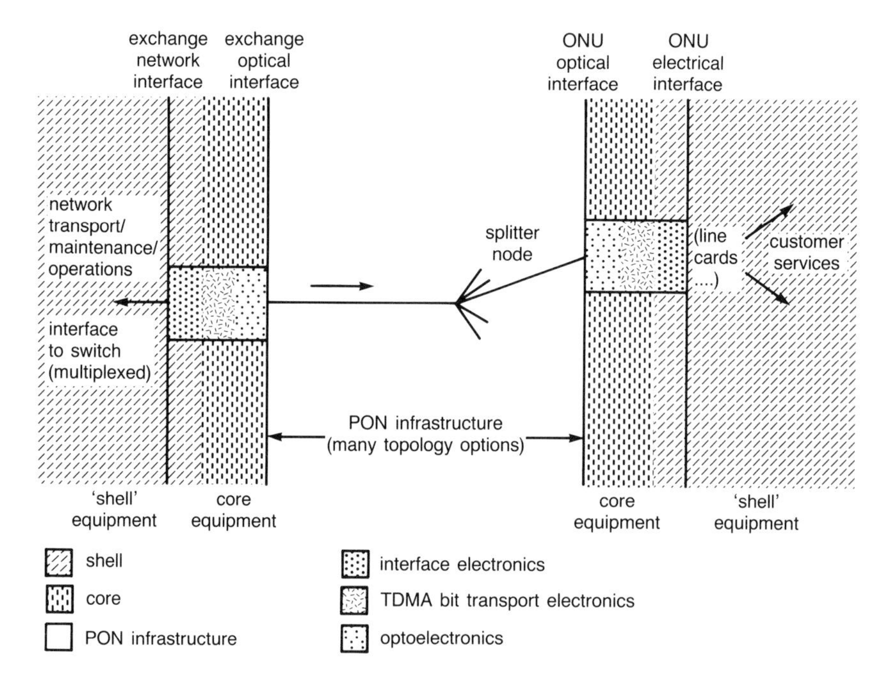

Fig. 15.6 Core/shell structure of PON system equipment.

15.3.5 Network evolution and longer term trends

15.3.5.1 Services

In the longer term it is expected that customers, whether business or residential, will increasingly request:

- a greater variety of services;
- improved quality of service coupled with lower costs;
- quick response times, in terms of provisioning and repair;
- management of bandwidth in more flexible and adaptive ways;
- services tailored to individual needs;
- a greater degree of direct customer control of network resources.

Thus the role of the optical fibre will be not just to provide scope for bandwidth expansion but to supply a much more flexible delivery mechanism for services than is possible with the traditional copper network. In this regard the intrinsic capability of optical systems (particularly PONs) to move traffic capacity dynamically around the network to track fluctuating customer demand, coupled with powerful in-built self-monitoring and diagnostic features, will be of major benefit in the future. Indeed the long-term vision could be for the provision of customer services in a manner analogous to the present-day supply of electricity — with the customer able to operate telecommunications appliances up to a maximum capacity rating within his premises, the network automatically dealing with the requirements of each active appliance and billing the customer strictly according to actual network usage.

In order to realize these much more responsive broadband networks it is important to recognize that other key network developments will need to include improved network-management techniques and network-based intelligence. Already there are large-scale investment and development programmes in progress aimed at achieving progressively advanced and flexible core transmission and switching underpinned by an end-to-end computerized management capability and distributed network intelligence.

15.3.5.2 Business customers

The broadband service requirements of business will tend to develop separately from those of the residential customer. Although the bulk of the requirements of business customers today are still provided over multiple connections to narrowband voice and data networks (with leased lines being used where organizations require a broadband — e.g. 2 Mbit/s — connection), the office environment is rapidly changing. Advanced office equipment and systems incorporating high-powered workstations and PCs are increasingly needing high-performance interconnection within buildings. LANs are satisfying this requirement and, to keep up with the pace of change,

LAN operating speeds are increasing from about 10 Mbit/s for Ethernets to LANs which will operate at 100 Mbit/s such as FDDI. The need to interconnect such LANs across public networks as well as other applications, such as high-definition graphics for CAD/CAM, bulk file transfers, video conferencing, electronic publishing, and image storage/retrieval, will be the key drivers for broadband business interconnect in the future.

15.3.5.3 Residential customers

The residential customer poses a significantly greater challenge than the business customer. On present cost estimates, extending FTTH for the single-line telephony customer is not yet viable. The best that can be currently contemplated is extending fibre as far as the final distribution point — e.g. as with the street TPON approach. As optical technology costs fall the cost differential between fibre and copper is expected to fall but it appears unlikely that FTTH will compete with copper for single-line telephony on a straight capital-cost-comparison basis.

Narrowband ISDN (2B + D, 144 kbit/s), which can be delivered on a conventional copper pair, looks set to meet the telephony and data service requirements for the domestic customer for some time to come. However, growth of narrowband services beyond a certain point (e.g. a second phone line, ISDN PC interconnect, home fax, hi-fi audio services) could trigger sufficiently increased revenues to justify fibre installation to the home. Early FTTH deployment is under active consideration in some countries, notably Japan and Germany. In the case of Japan the aim is to develop new narrowband services as a stepping stone to broadband (see Chapter 1). In Germany, the need to renovate the old network in the east of Germany may create a unique opportunity to deploy fibre as (virtually) the entire duct infrastructure is replaced.

Otherwise FTTH will need to await the demand for entertainment TV services. In many countries these may need to be differentiated in quality from those already offered by existing satellite or coaxial cable TV networks. Potential services are cable TV, pay TV, interactive TV, multimedia-based home shopping [10], high definition television, video on demand, sophisticated computer games with high-quality graphics, 'virtual reality' [11] based services, etc. The more speculative possibilities in this list will be critically dependent on future entrepreneurial and creative developments triggered by the widespread availability of an initial entry-level fibre infrastructure.

Although many network trials have been carried out to deliver broadband services to residential customers, costs to date have been too high to justify wide-scale deployment. Regulatory restrictions and the relative immaturity of the public demand for broadband services have also acted as a considerable discouragement to progress to date. Looking to the future, the more modern FITL architectures (see Chapter 4), coupled with developments in optical technology, offer a better prospect for economic viability. In particular, these networks are potentially able to operate cost-effectively at lower levels of initial service penetration.

15.3.5.4 Longer-term technology trends

For the telephony entry networks such as TPON, existing device technologies will ultimately restrict the range of applications to which fibre can be economically applied and further technological developments will be needed to increase network penetration. Fundamentally lower-cost device technologies, coupled with new system design approaches, should allow very small ONUs to become viable and bring closer the deployment of fibre-to-the-home (FTTH). Recent work, for example, has investigated the potential for the use of a high-performance single-chip optical transceiver to replace four optical/optoelectronic components as used in a conventional duplex PON design (Fig. 15.7) [12]. The transceiver is based on a special semiconductor laser and has demonstrated a PON power budget performance that approaches, within a few dBs, the performance of a conventionally designed system. A 90% efficient bit transport protocol has been devised to govern the operation of the transceivers over a PON.

In the longer term active double-star networks for FTTH (uprated from FTTC telephony entry designs being trialled in North America) are intended to move towards the use of digital feeders (to the cabinet or remote node) operating at the Gbit/s rate with 600 Mbit/s digital links to the customer to cater for multiple, switched TV/HDTV channels to the home. Asynchronous transfer mode [13] could emerge as the preferred transport format for residential and business customers although this is controversial and by no means guaranteed. Cabinet-mounted active nodes could become displaced to the exchange if direct point-to-point links from the exchange to customer become economic.

In the case of PONs, higher capacities will increasingly be required to meet the requirements of business customers and could also, in the longer term, migrate towards an ATM-based protocol for integrated service delivery. Experimental ATM PON networks, such as BT's APON and CNET's SAMPAN (see Chapter 12 and Fig. 15.8) [14] have been reported and these will demand improved performance, but still low-cost, optical devices if similar levels of network split are to be maintained. The APON network permits the standard BISDN 155 Mbit/s ATM cell stream to be shared

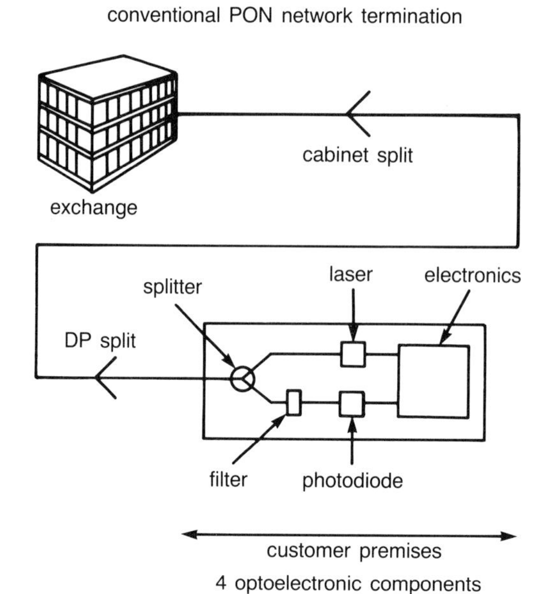

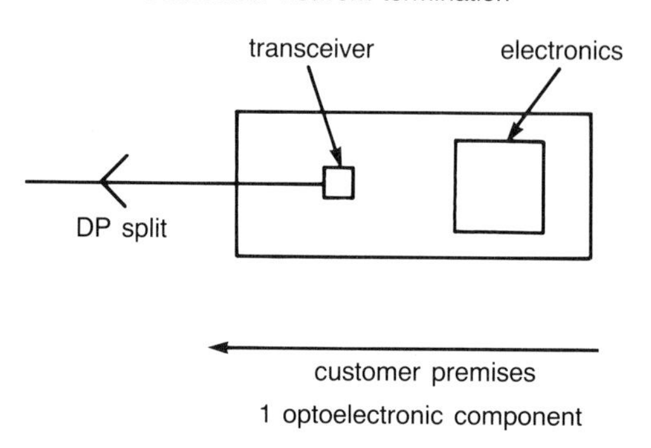

Fig. 15.7 Conventional PON network termination.

amongst up to 64 customers on a PON — thus offering a low-cost, shared-access method for the introduction of advanced service capability to smaller business and, eventually, residential customers.

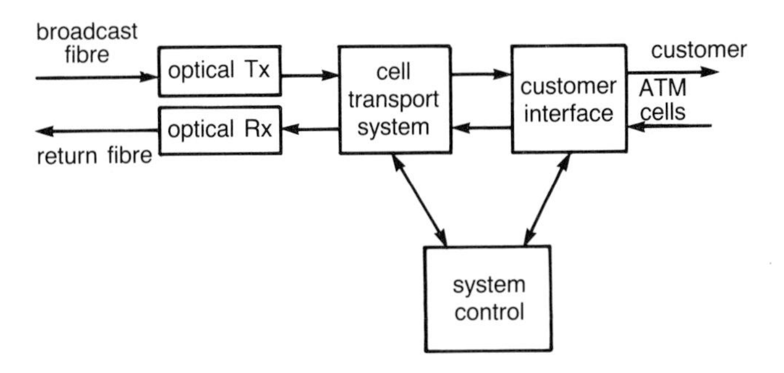

Fig. 15.8 Customer end of APON.

For both the residential and business sectors, the ability to upgrade PONs to provide a wide variety of multivendor network services will be much enhanced by the availability of low-cost wavelength division multiplexing technologies that can permit the customer to select from a range of incoming wavelengths. Technology at the ONU will again be crucial to achieve the higher functionality required at low cost. As an example of the advanced technologies now beginning to emerge in this area, Fig. 15.9 illustrates a four-channel (wavelength) optical array receiver, with individually programmable channel sensitivities, developed recently at BT Laboratories [15].

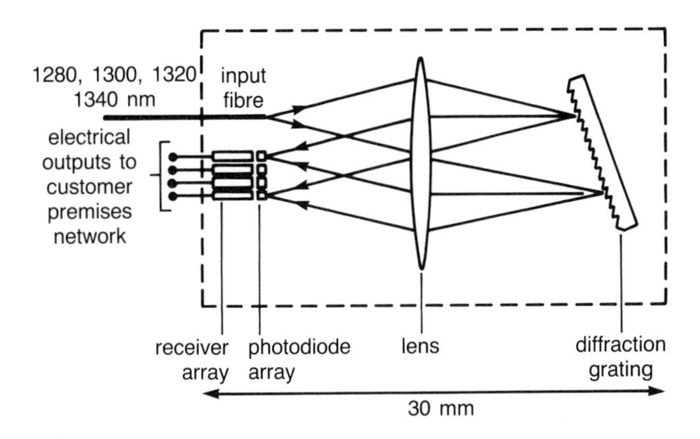

Fig. 15.9 Hybrid demultiplexer/receiver.

Eventually it will become possible to transmit hundreds of optical wavelengths over the network leading to a vast choice in network services and service providers. Already technology demonstrating tens of optical wavelengths is beginning to emerge from the laboratory. Technical breakthroughs on optical amplifiers are also of importance, allowing large splitting ratios to be achieved and services to be delivered from a small number of very distant, 'local' exchanges. A recent experiment at BT Laboratories, for example, demonstrated the feasibility of broadcasting 384 TV channels to 39.5 million customers from a single headend via only two stages of optical amplification (Fig. 15.10) [16].

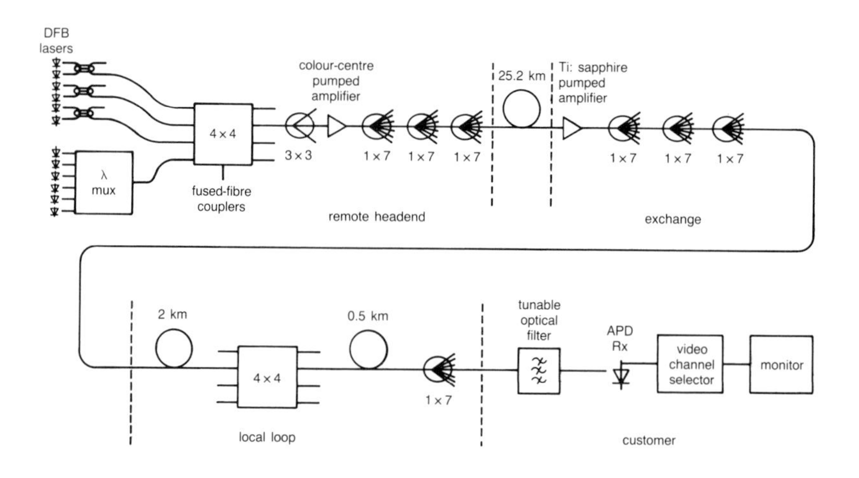

Fig. 15.10 Broadcasting 384 video channels to 39 530 064 customers, using two stages of erbium-doped fibre amplifiers.

Optical switching will also play a part in future networks. The emergence of tunable optical devices (lasers and filters) will lead to distributed forms of switching that will blur the boundaries between the local network, the exchange switch and the core network. In time the network could come to resemble an optical ether where optical signals are transparently routed from end to end in a manner analogous to present-day radio communications.

Major challenges face the optical researcher in making these futuristic network possibilities a reality. Viable optoelectronic integrated optical circuit (OEIC) technologies will be particularly important so that the more complex optical 'circuits' required can be realized at acceptable cost.

15.4 COPPER DEVELOPMENTS

However aggressively fibre is deployed the copper pair will still be the dominant carrier in the local loop well into the next century. Key issues for telecommunications companies are therefore to improve the quality of service that can be offered over the present network and to raise the efficiency with which the vast embedded base of copper is used. Particular developments in train are improvements in line surveillance, so that preventive maintenance can be scheduled ahead of actual service failure, and the application of modern digital signal processing techniques (see Chapter 5) allowing the copper pair to carry higher-capacity signals. These advances in copper technology will make the optical fibre work still harder in the future to establish itself, particularly in certain network situations. High-rate digital subscriber line (HDSL) technology will in the future allow broadband local access at rates up to 2 Mbit/s over unscreened copper pairs. Essentially in these systems, adaptive techniques are used to correct for bandwidth and crosstalk limitations. HDSL systems can be used for pair gain, where additional capacity is added to an already working pair. Various additional services such as ISDN, private circuits and low-definition video telephony can also be supported as well as standard telephony.

15.5 RADIO IN THE LOOP

A further factor of increasing importance for the future local loop is the development of mobile communications based on radio techniques. In particular a move can be expected towards a personal communications service which is much more widely available and lower cost than the present national cellular radio systems [17]. It is even possible in the longer term that the hand-held personal communicator may become the prime method of telephony access for many customers.

Such developments will depend heavily on the establishment of sufficient intelligence, fast signalling and powerful databases within the core of the network to allow users to be traced regardless of location. Within the access network greater frequency allocation and much higher efficiency in usage of the radio spectrum will be necessary to cope with the greater customer densities involved. This will lead to higher-frequency operation (typically around 2 GHz) and also to smaller sizes of radio cells (around 100 m) to allow greater reuse of available channels. Figure 15.11 illustrates the likely layout for such cordless access systems. The radio DP could be mounted on a dedicated pole or on street lamps or other prominent objects. Transmission links to the DP could be by copper, fibre or, in low customer-density rural areas, fixed radio links. Recent work [18] has demonstrated the potential advantages of using fibre links which transport the radio-frequency signals direct to the DP. This allows the radio antenna to remain at the DP but relocates much of the electronic equipment at the local exchange — where further economies can be achieved by sharing equipment amongst several radio DPs.

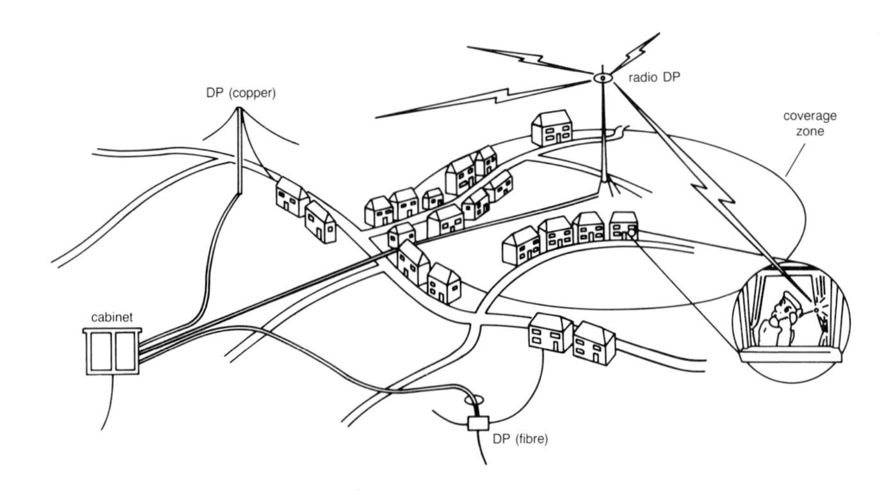

Fig. 15.11 Cordless access.

15.5.1 The shape of the future local loop

The exact blend of copper, radio and fibre in the future local network will be dependent on many factors, some technical and cost-related, but some related to even less predictable factors such as the evolving regulatory climate, the emergence of new services that grab the public imagination and the degree to which mobile telephony spreads beyond the business sector to all walks of life. Against this background, predictions on the shape of the future local loop are hard to make. Nevertheless certain trends are becoming clearly established.

First the strategy for the introduction of optical fibre in the loop is no longer pinned to the simplistic and visionary 'broadband wired city' ideas of earlier times. Although evolution to that end point is still in view, it is now widely accepted that a more pragmatic approach of introducing fibre against known services is essential if progress is to be made. In most countries this means fibre for telephony with broadband coming later. Opinions on timescales for the latter vary markedly depending on the prevailing regulatory climate. In much of Europe (e.g. Germany, France, Holland) the PTT has an effective monopoly of cable television delivery and views are relatively optimistic on the need for early broadband upgrade. In the USA and Japan the regional Bell operating companies and NTT respectively are not currently allowed to deliver such services. In the UK the British government has indicated that BT will not be allowed into this market for some years.

Countries with more favourable prospects for entertainment TV may give greater emphasis to installing FTTC systems for telephony in residential areas so that they can be available for broadband upgrade at a relatively early date. In the UK there is likely to be a more cautious approach, with FTTC installed only where the economics for supplying existing services are advantageous and once the full cost implications of installing large numbers of active DPs in the network are fully understood. Cordless access, using radio drops from FTTC systems, is a potential contender for providing the final link to suburban residential customers — as well as the copper pair. In rural areas copper and fixed radio links, in certain situations, are likely to remain the most cost-effective carriers for many years to come.

In the UK, fibre will be more aggressively deployed to the business sector than to the domestic customer for the foreseeable future. BT has already established a policy of deploying point-to-point fibre systems to large business customers. The next step is to target FTTO for small and medium business customers in new development areas using business TPON (or similar) systems. It is hoped that large-scale deployment of such systems will become possible in 3-4 years' time.

A major objective of many telecommunications companies is to adopt a 'cap copper growth' strategy as soon as possible in order to minimize investment in new copper that could become obsolete within the normal timescale for asset depreciation. A further desire is to avoid the expensive new duct schemes that would be needed to house large amounts of new copper. In order to move in this direction 'fibre-ready' policies are being considered whereby the local loop will be prepared for fibre as part of standard working practices during normal (copper) service provision. This will involve the installation of fibre spines in the network from exchange down to street-cabinet level. Duct space will also be reserved between the cabinet and the customer, whenever possible, by installing blown fibre tubing (see Chapter 11), into which fibre can be easily inserted later when required.

These steps will begin to prepare the network for more general fibre deployment when this can be economically justified. The gradual movement of business customers from copper to fibre should also greatly help the problem of coping with the large element of churn in business circuits, the greater flexibility of fibre systems allowing capacity to be moved more easily around the network to track customers' requirements, while fewer copper pairs need to be held in the network to meet unpredictable business demand, leading to a more efficient use of the embedded copper base. Many existing copper pairs should be freed to help meet growth where fibre is not practical. Copper-pair gain technology will also aid the movement towards 'cap copper growth', increasing pair supply in areas of shortage.

Ultimately, an all-fibre network must be the aim in order to achieve the potential for virtually unlimited growth in new and innovative services. Realistically, however, this may take many decades to occur with progress much influenced by the pace at which service developments can actually be successfully pioneered. This may lead to 'interim' stages in network development, each of which could last for some time, with progression to the next stage dependent on prevailing market and regulatory circumstances. One possible 'interim' stage could be the emergence of the concept of the universal hub (see Fig. 15.12). Thus current trends in the cable TV industry [19] are towards a mini-hub concept with fibre feeders from a remote head-end down to a flexibility point close to the customer. The likely deployment of FTTC fibre systems as a first step in residential districts has already been described. A third type of fibre hub could be the node for a radio cell. It is possible therefore that a common fibre/hub technology could emerge capable of feeding each type of final drop, depending on the particular type

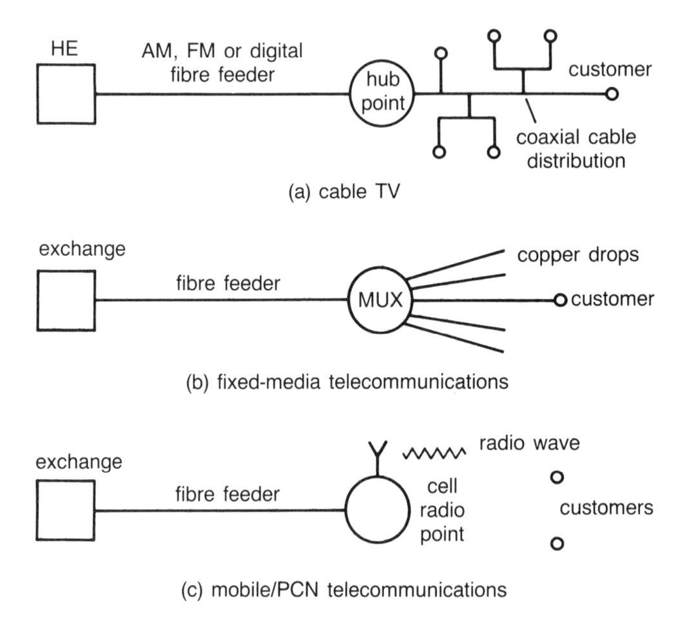

Fig. 15.12 Universal hub concept.

of output card fitted within the hub node. A problem could be the different sizing of nodes required by the different customer ‘drop’ technologies. However, this might be overcome by designing with a suitably modular approach. The likelihood of this scenario emerging as a realistic option may be chiefly dependent upon how practical and cost-effective the deployment of large numbers of active nodes in the network turns out to be — as opposed to how rapidly the market situation develops to justify full FTTH deployment.

15.6 CONCLUSIONS

The local access network, largely unchanged for many decades, is now on the threshold of major technological changes which will lead to a dramatic increase in the range and flexibility of services, both narrowband and broadband, that can be offered to customers. Key technological developments will be the progressive deployment of optical fibres coupled with the complementary use of radio techniques and high-rate digital transmission over copper-pair cables.

Early in the next century, many business customers, including shops, local banks and small industrial units, could be connected by fibre carrying LAN-LAN interconnect, fast file transfers, imaging and video conferencing. Penetration of fibre-to-the-home will be slower, depending on market and regulatory circumstances, and will occur only for customers subscribing to new types of service. Fibre may be deployed down to the final distribution point (DP) in the local network, some tens of metres from the customer, at a much earlier stage.

Realistically, copper will remain the dominant carrier in the residential local network for many years to come — given the long lead times needed to install fibre to millions and millions of customers — even once market demand has been clearly established. Distributed entertainment TV service, enhanced quality TV, video on demand and interactive video services are the most likely applications to motivate the installation of fibre to the home although the marketing of multiple narrowband services cannot be ruled out.

The deployment of advanced technology in the local network will lead to major improvements in network flexibility and quality of service. The replacement of large copper cables with low fibre-count cables will improve network repair times and encourage the development of robust, self-healing networks based on alternative routeing techniques.

In the short term the crucial issue for local loop fibre systems is to achieve cost parity against copper for the provision of existing narrowband (telephony and data) services. Progress on standards and international convergence is now essential if truly wide-scale deployment is to be achieved. In the long term an all-fibre network, enhanced by future generations of optical and optoelectronic technologies, can provide the basis for virtually unlimited service growth.

REFERENCES

1. Dufour I G: 'Flexible access systems', BT Eng J, 7, pp 233-236 (1989).

2. Hoppitt C E and Rawson J W D: 'The United Kingdom trial of fibre in the loop', BT Eng J, 10, pp 48-58 (April 1991).

3. Shumate P W and Snelling R K: 'Evolution of fiber in the residential loop plant', IEEE Communications Magazine, 21st Century Subscriber Loop Issue, 29, No 3, pp 68-74 (March 1991).

4. Abiven J: 'Transmission of narrowband services on a passive optical bus', EFOC/LAN, Munich, pp 387-391 (1990).

5. Tromp H R C, Nijnuis Y, Boomsma Y and Bakker J: 'Fibre-to-the-home in the Netherlands', International Symposium on Subscriber Loops and Services, Amsterdam, pp 253-259 (April 1991).

6. RVS: 'Raynet video system', Technical Description, Raynet Corporation, Menlo Park, Ca, USA (1991).

7. Kimura Y, Suzuki K and Nakazawa M: '46.5 dB gain in Er3+-doped fibre amplifier pumped by 1.48 μm GaInAsP laser diodes', Electron Lett, 25, No 24, pp 1656-1657 (November 1989).

8. Lobbett R A et al: 'System characterisation of high gain and high saturated output power, Pr^{3+}-doped fluorozirconate fibre amplifier at 1300 nm', Electron Lett, 27, No 16, pp 1472-1474 (August 1991).

9. Bellcore Technical Advisory Document, TA909 (December 1990).

10. Sawyer W D M, Salladay J H and Snider M W: 'The gateway role in an open-access, high-capacity network', International Symposium on Subscriber Loops and Services, Amsterdam, pp 401-407 (April 1991).

11. Rheingold H: 'Virtual Reality', Secker and Warburg (1991).

12. Smith P: 'A high performance optical transceiver for low cost fibre-to-the-home', EFOC/LAN 91, pp 331-337 (June 1991).

13. Boulter R A: 'Access to a broadband ISDN', BT Technol J, 9, No 2, pp 20-25 (April 1991).

14. Du Chaffaut G: 'An ATM cell based transmission system on a PON structure', Globecom 90, San Diego, paper 303—2 (1990).

15. Griffiths D C et al: 'A hybrid integrated four channel wavelength demultiplexing receiver', European Conference on Optical Fibre Communication, Paris (September 1991).

16. Hill A M, Wyatt R, Massicott J F, Blyth K J, Forrester D S, Lobbett R A, Smith P J and Payne D B: '39.5 million way WDM broadcast network employing two stages of erbium-doped fibre amplifiers', Electron Lett, 26, No 22, pp 1882-1883 (25 October 1990).

17. BT Technol J, 8, No 1, Special Issue on Mobile Communications (Jan 1990).

18. Cooper A J: 'Fibre/radio for the provision of cordless/mobile telephony services in the access network', Elect Lett, 26, No 24, pp 2054-2056 (Nov 1990).

19. Session on 'Fibre towards the cable TV home', 17th International Television Symposium, Montreux, Cable TV sessions record, pp 501-617 (June 1991).

Index

Page numbers in italic represent tables, those in bold represent figures.